CROWOOD METALWORKING GUIDES

MAKING AND USING SMALL WORKSHOP TOOLS

NEIL M. WYATT

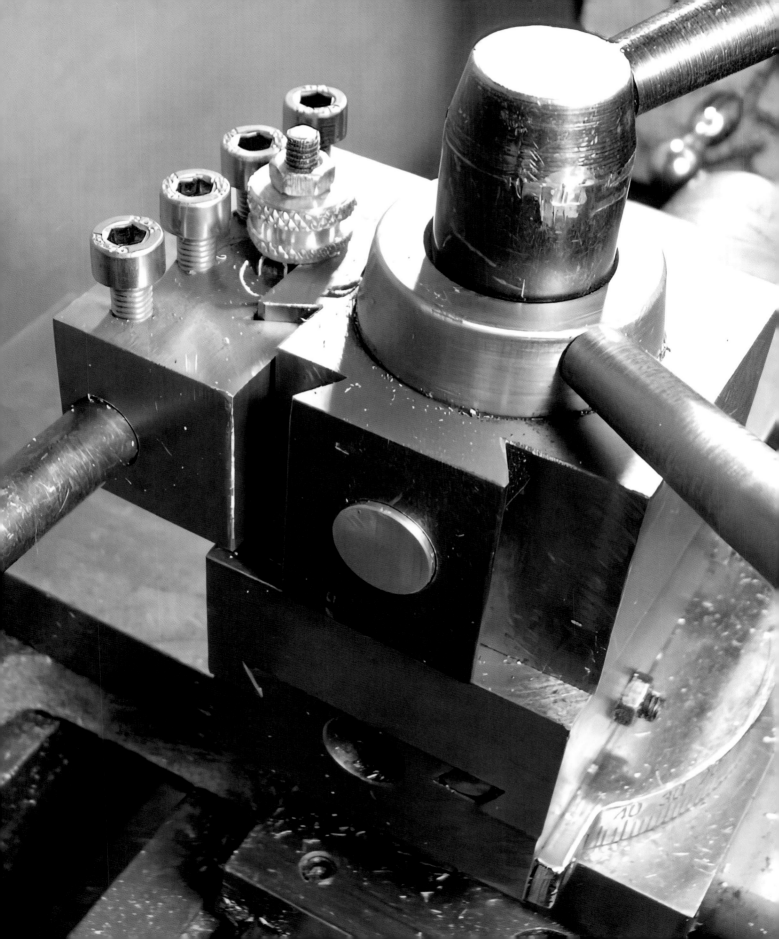

CROWOOD METALWORKING GUIDES

MAKING AND USING SMALL WORKSHOP TOOLS

NEIL M. WYATT

THE CROWOOD PRESS

First published in 2022 by
The Crowood Press Ltd
Ramsbury, Marlborough
Wiltshire SN8 2HR

enquiries@crowood.com
www.crowood.com

© Neil Wyatt 2022

British Library Cataloguing-in-Publication Data
A catalogue record for this book is available from the British Library.

ISBN 978 0 7198 4143 9

Typeset by Simon and Sons
Cover design by Maggie Mellett
Printed and bound in India by Parksons Graphics

Contents

Introduction

I look back and realize that I have been participating in model engineering as a serious hobby for almost a quarter of a century, and enjoying being in and around workshops and tools for over twice as long. Yet, I recognize that I have more to learn than I will ever know about the subject. Even so, it is impossible to have hobby engineering as such a significant part of your life for so long without gathering a great deal of practical experience and useful knowledge. This book is intended to pass on some of the advice and ideas that I feel would have been useful when I first started out in this hobby. It is also a chance to bring together some of the small tools, jigs and accessories I have made in my own workshop.

In the past, books on smaller workshop tooling have focused solely on a series of shop-made gadgets and devices. Having been encouraged by the writings of Tubal Cain (Tom Walshaw), Stan Bray, Ivan Law, George H. Thomas and others, I have no doubt that such project-based ideas are really useful, not only as a practical guide to making useful devices but also for their insights and inspiration for designing and making my own workshop tooling. For that reason, this book includes many such project ideas at all levels, from the very simple to the fairly complex.

My aim has not been to provide every last detail on each example, as that could become very repetitive. Instead, most of the complex devices, such as the boring head and box tool, are fully described. Other ideas, such as the toolpost spindle based on a power drill gearbox and micrometer stand, are described in outline. They are intended to inspire you to go through your own 'scrap box' and work with what you have available.

What this approach misses, however, is the opportunity to pass on useful guidance on using and adapting commercial tooling. Usually, such advice appears incidentally in articles on some other subject, but in this book, I have tried to be more systematic. I have set out to pass on as much good, general advice of this kind as I can fit into one volume. It should help you make the most of your hand tools and small accessories, while equipping you with the ideas and knowledge to make many of your own.

It has been hard to decide what to cover and what to leave out because of it being too straightforward. I have decided to avoid items such as screwdrivers and I also abandoned a lengthy review of different types of hammer, but I have included many less intuitive tools. I have also gone into some detail on files, as I feel their potential and variety are poorly understood by many beginners.

I could have explored many useful workshop techniques, such as brazing and anodizing, but these would have strayed too far from the core subject. These pages do have a more detailed explanation of hardening and tempering carbon steels because these skills are essential for toolmaking, particularly when making specialist cutters of various kinds.

As a result of my approach, this book is a true miscellany, and no doubt every reader will find different parts more useful than others. Whatever you do in your workshop, and whatever your level of skill and knowledge, thank you for your interest in my thoughts and experiences. I hope that you find useful and interesting ideas in this book that will help you in your workshop journey.

Neil Wyatt

A Word on Safety

Some of the projects in this book involve the use of machine tools, which can of course present a safety risk. Even hand tools can be the cause of serious accidents if used unwisely. Below are some pointers towards the safe use of all workshop tools. Always remember that safety is your responsibility; if you are not sure about something, check with someone who is.

SAFETY CONSIDERATIONS

Electrics

As with any electrical equipment, care must be taken with leads and around the electrical parts of the machine, bearing in mind that tools move, and wires can be drawn into machinery. Ensure that plugs are wired correctly and that any fuses that blow are replaced with an item of the same type and value. If you have an electrical problem or you find your fuses blowing repeatedly, consult an electrician. Do not attempt to dismantle or repair any electrical elements of the machine unless you are competent to do so. If you use cutting fluid of any kind, take great care to make sure it goes nowhere near the motor, control box or wiring harness.

Emergency Stop

Make sure you are familiar with the emergency stop button on power tools.

The more familiar you are with using one, the faster you will be able to hit it when you need to.

PPE

Most small tools do not demand much in the way of personal protective equipment, although the Health and Safety Executive (HSE) recommends that eye protection is worn when using a lathe or mill. When cutting something like brass, which can send a vertical shower of small chips into the air, full eye protection is essential. For most turning, many people find a pair of safety spectacles or goggles will be sufficient, but bear in mind that flying swarf is not

the only risk. It is also possible for loose workpieces to fly out of the chuck at high speed. When using hand tools, use your judgement. For rotary tools, eye protection is essential, while hammers and chisels can cause chips to fly.

Guards

While modern machines are generally fitted with good guards, older machines are often equipped with no guard at all, or with a tiny guard that serves little practical purpose. The worst examples offer no real protection in the actual working area and in some cases they even limit the usability of the lathe. If a guard interferes with the use

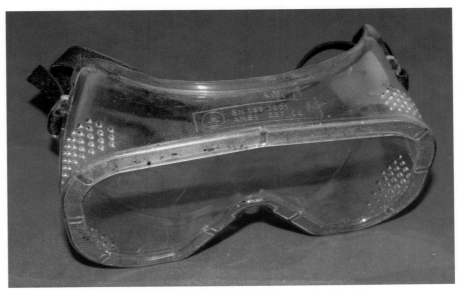

Fig. Safety.1 Safety googles are essential for eye protection.

Fig. Safety.2 A simple swarf guard from polycarbonate sheet mounted on a magnetic block.

velcro or snap fasteners at the front. Be careful to keep long hair tied back, too.

Do not use cloth rags near moving work. Kitchen roll is much better as, if caught, it will simply rip and not snag.

Lighting

It's always important to ensure the work area is well lit. Ideally, use low-voltage lighting and ensure that any luminaires are proofed against the entry of swarf or liquids. Avoid old-fashioned fluorescent lights, as these can give the impression that the lathe is not rotating if the machine speed synchronizes with the mains frequency.

Noise

Most machine tools are not excessively noisy. If yours seems to be louder than usual, it is probably trying to tell you things are not right; for example, a cut may be too heavy, causing shrieking chatter, or an interrupted cut may have been taken too quickly or aggressively. There may be times, however, when you cannot stop a loud or unpleasant sound from a particular tool due to some resonance or other. For these occasions, you can keep a pair of ear protectors handy, but a pack of cheap earplugs is cheaper and equally effective. Do take care, though, not to be caught unawares by someone entering the workshop when you are wearing them.

Organization

In the real world, a workshop will not be as clean and tidy as an operating theatre, but that does not mean that you should

of a machine, the user may be inclined to leave it open, which means that it is pointless.

I have found that a sheet of polycarbonate about 200–250mm (8–10in) square, mounted on an arm fixed to a weighted or magnetic base, can make a guard that is easy to use and effective. This can be easily cleaned and placed so that it protects the operator without interfering with the use of the machine.

Chemical Fluids

It may be a surprise to many, but the HSE has found that health issues such as dermatitis, asthma and lung damage from cutting fluid represent one of the biggest hazards associated with machining. Most people do not use floods of coolant – indeed, many do most of their work dry or just use small amounts of neat cutting oil applied with a brush or dribbled on to the work. This is less hazardous, but the fluid can still get thrown around or give off unpleasant fumes if the work gets hot. Take care to make sure you do not breathe in any fumes or overspray, and consider using barrier cream and/or a face mask.

Gloves and Other Clothing

Do not wear gloves when working with a machine like a lathe, mill or drill. If they are at all loose, they can be pulled into a moving machine, leading to terrible injuries.

It is unlikely that many people would consider wearing a tie in the workshop these days, but it is vital to be aware of any loose clothing. A more modern hazard are the drawstrings found on a hoodie, which can dangle right over the work as you lean forward for a closer look. The best workshop clothing is an overall with

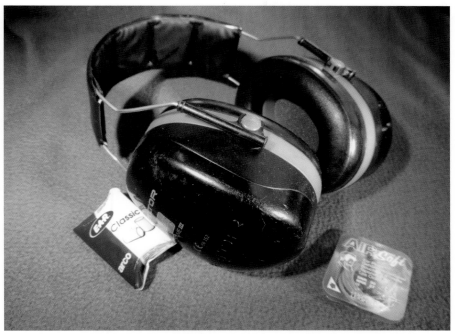

Fig. Safety.3 Ear plugs and ear protectors are both effective, so use what you find most comfortable.

work in the midst of chaos. Make sure there is plenty of space for you to move around and that there are no trip hazards waiting to catch you out when your attention is elsewhere. Keep wires and other tools out of the working area and try to avoid anything that might tempt you to lean over a moving machine.

Children

Responsible youngsters should not be discouraged from entering the work-shop, but they must be properly super-vised at all times. Make sure you take extra care for their safety and be mindful of the fact that any task you give them should be within their capabilities.

SOME DO'S AND DON'TS

◆ **Do** give machining your full atten-tion. Avoid distractions in the work-shop. You will not do good work if you cannot focus on it and the majority of accidents are due to inattention.

◆ **Do** remember that you are not the only person who may walk into your workshop and switch on the power, so make sure things are left in a safe state.
◆ **Do** use eye protection and suitable guards.
◆ **Do** take good care of your tools.
◆ **Don't**, ever, leave chuck keys in a chuck.
◆ **Don't** leave machines switched on when setting up work or changing tools.
◆ **Don't** try to remove swarf by hand. It is sharp and can easily cut or pull fingers into the work. Keep a bit of wire coathanger with an 'L' on the end handy, for use as a swarf 'puller and poker'.
◆ **Don't** make electrical modifications or repairs unless you have the skills and experience required.
◆ **Don't** leave unattended machines switched on.

Safety is ultimately a matter of common sense and paying attention. Your eyes, ears and fingers are your most valuable assets, so take the same care of them as you would of any other irreplaceable tool.

1 Files and Filing

It seems appropriate to start by looking at files – some of the simplest and yet often most frequently misunderstood or abused of small tools. Files are hand tools with flat or curved surfaces bearing a series of fine teeth for the removal of metal. They usually have a pointed 'tang' at one end for fitting a handle. Ancient civilizations understood various ways of using abrasives, such as sand embedded in wood, and bronze rasps with many small teeth were known of in ancient Egypt and elsewhere. Metal files such as those that would be familiar today did not appear until the Middle Ages, when the means of making them was often a closely guarded secret.

Modern files typically are made of hardened steel and have regular teeth made by one or two sets of angle grooves. They come in various shapes, sizes and grades or 'cuts', from dead smooth through smooth to bastard cut, which is the coarsest. Files made by alternative methods of manufacture, such as the bonding of diamond or carbide particles to the surface, are increasingly commonly available.

FILING TECHNIQUE

The efficient and accurate removal of metal using a file is a technique that approaches an art. Good filing technique is an invaluable asset. In the past, it was a traditional first duty for any apprentice to take an irregular lump of metal and file it into some precise shape.

For large files, the best technique is to hold the workpiece in a vice, at elbow height. Traditionally, the file should be pushed gently forwards across the work, with one hand on the handle and the other steadying the far end. This is as much about accuracy and steady technique as applying extra force, and even small files can benefit from a steady finger pressure on their tip. Typical filing rates are around sixty to a hundred strokes per minute, and it is considered good practice to lift the file from the work, or at least remove all pressure, on the return stroke. Try to keep the file level, as otherwise the corners of the work will be rounded, or you may even end up with a convex surface.

When aiming to size an object accurately it helps to have a scribed line or a witness mark – even a line drawn with a Sharpie – to work to. Strokes should be concentrated at the highest points but it is important to vary the position of the file to avoid making grooves in the work.

Draw filing is a useful technique for finishing flat surfaces, involving sideways strokes of a file held sideways with a hand at each end. It creates a similar texture to a surface finished by grinding, and is an excellent way to tidy up the sawn edges of steel plate. Draw filing is much less aggressive than ordinary filing and can be a useful technique when trying to achieve a close fit by hand.

Other uses for files include making keyways or notches. Fig. 1.4 shows a jig that can be inserted in a small gear or flywheel to guide the filing of a narrow keyway. It is just a simple top-hat shape that is a good fit in the bore with a filed or milled slot to guide the file. Opinions vary on whether it is better to harden

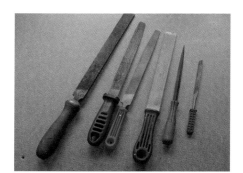

Fig. 1.1 A selection of different file types.

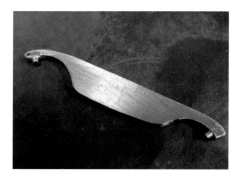

Fig. 1.2 Shop-made C-spanner with draw-filed finish.

Fig. 1.3 Bar in vice for draw filing.

Fig. 1.5 A quality handle for larger files.

FILE CUTS AND TYPES

Below is a broad overview of the commonest file cuts and their typical applications:

◆ Bastard-cut: the coarsest grade of metal file, used for rough shaping and the rapid removal of metal.
◆ Second-cut: with moderately coarse teeth, a good general-purpose file for metal removal and shaping. It is not suitable for fine finishing, although it can give a reasonable appearance when used for draw filing.
◆ Smooth: a cut with small teeth, giving a smooth surface finish.
◆ Dead smooth: a file with very small teeth, giving a fine surface finish.

such a jig or not, but if it is only for a single use, then it is probably best not to harden. It is always handy to have a file handy for tidying up the sharp corner or to 'break the arris' where two machined surfaces meet by creating a small chamfer. This is often done by draw filing along (not across) the edge as an alternative to a deburring tool drawn along the corner of the work. Such corners hold paint better as well as being more pleasant to handle. Care should be taken on scale models not to create an out-of-proportion effect by being too enthusiastic.

Files should have a well-fitting and comfortable handle fitted to their tang; not only is this essential to prevent injury on the tang, but it is also an essential aid to easy and accurate use. The smallest of files (those with round shanks) may be used without a handle, but these too will usually benefit from having one fitted.

A filing rest is a height-adjustable jig with two rollers used to support a file when it is being employed to put accurate flats on an object held in a stationary lathe. It usually has a raised ridge, to prevent the file making contact with and damaging the chuck.

Over time, files may clog up, especially when used on softer metals such as aluminium and its alloys. A brass wire brush or file card (a handle-less brush especially for this purpose) worked along the file's grooves can bring back much of the file's effectiveness. Another way to clean a file is to rub a hard brass rod across the grooves – it will rapidly adopt a shape that reaches into the grooves and clears the clogging.

Fig. 1.7 Various file cuts (L to R): Millennicut with 'chip breaker' notches, curved, bastard, medium, fine and smooth.

Fig. 1.4 Slotted 'top hat' jig to guide filing of keyways.

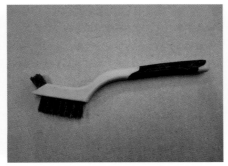

Fig. 1.6 A stiff brass brush is an alternative to a file card.

Fig. 1.8 Examples of file shapes (L to R): round, crossing (bi-convex), half-round, flat, warding (flat and tapered), knife, three-square.

It is not suitable for the removal of large quantities of material.

◆ Single-cut: with a single series of teeth at an angle of about 25° to its length, this sort of file typically gives a smoother finish than a double-cut file of similar tooth size.

◆ Millennicut: a brand that supplies a single-cut file with very sharp, high-quality milled teeth that have 'chip breaker' grooves cut across them. Millennicut files are good for filing softer metals, such as pure aluminium, and even body filler, without clogging.

Files also come in a wide range of shapes, to suit various purposes, the commonest are:

◆ Flat: the basic rectangular cross-section shape of general-purpose file. It may be tapered at the end. Flat files often have a 'safe edge' – an edge with no teeth on it – allowing for working up to an internal corner without marking it. Before using a file in this way, it should be carefully inspected, as there may be burrs that will spoil the safe edge; if this is the case, they can be carefully ground or stoned off.

◆ Hand: very much like a flat file but tapered in thickness towards the tip.

◆ Square: a file with a square cross-section.

◆ Three-square: a file with an equilateral triangular cross-section.

◆ Half-round: a file with a cross-section convex on one side and flat on the other. The rounded side is used for filing concave shapes.

◆ Double convex: a file with an oval-shaped cross-section.

◆ Double-cut: a file with a second series of teeth cut across the first series.

◆ Round: a file with a circular cross-section.

◆ Rat tail: a round, tapered file with a circular cross-section, often rather coarse cut and used for opening out holes or filing notches.

◆ Knife: a file with a long, thin triangular cross-section.

◆ Taper: a file that reduces in width towards the tip, typically over around a third of its length.

As well as these more common types, there are also many options for specialist purposes. There are too many to list in full here, but they include rasps, which have individually cut teeth (normally coarse) and are typically used for softer materials such as wood or plastics. Also available are needle or Swiss files, which are small – typically 150–200mm (6–8in) long – and normally with a relatively fine cut. Such files are available individually and as sets in a wide range of different profiles. Quality is reflected in cost and the best single files will be more expensive than a whole set of cheap files. The name reflects the use of such files (and smaller) in Switzerland's traditional clock- and watch-making industry. They are often supplied without handles. If the tang is rod-shaped rather than sharp, they could be used for light hand work without a handle but should never be used in this way near machinery. Push-fit handles are available to fit, some of which have a collet so they may be swapped between files.

Carbide files are suitable for use on tough materials. The cutting surfaces are covered with small pieces of tungsten carbide bonded to the surface, which function as teeth. These files are ideal for tidying up iron castings, which often have a tough 'skin'. Diamond files similarly use tiny industrial gemstones bonded to a carrier and have much finer cuts. Diamond-encrusted 'cards' and 'slips' are particularly useful for tasks such as honing the edge on high-speed steel (HSS) or even carbide-tipped tooling.

An abrafile is a long, thin file blade designed to be used under tension, so it is normally used fitted in a hacksaw, junior hacksaw or coping saw frame according to size (Fig. 1.14). For cutting rather than filing, it can cut in any direction and is ideal for cutting out complex shapes from sheet metal. Abrafiles are

Fig. 1.9 Small files (from top): round file in handle, two diamond-coated files with dipped handles, two Swiss files with miniature handles.

Fig. 1.10 Carbide-coated file.

becoming hard to source; blades that are essentially rods coated in tungsten carbide chips are generally easier to find.

Riffler files are double-ended tools with a small, shaped file at each end. They may be dramatically curved or tapered and offer a multitude of solutions for filing delicate and unusual shapes. One very handy application is 'spot filing' a raised area on a flat surface using a convex riffler.

Fig. 1.14 Carbide file in a hacksaw-like frame for tile cutting.

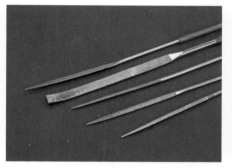

Fig. 1.11 Diamond-coated files: note the flat file bent up, so it can function as a riffler for spot filing.

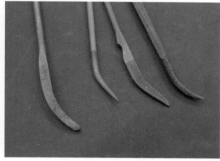

Fig. 1.15 Examples of riffler file shapes.

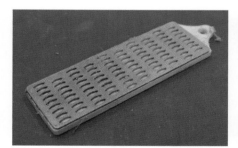

Fig. 1.12 Plastic-backed diamond slip for honing tools.

Fig. 1.16 These model locomotive coupling hooks were hand-filed from flat blanks.

Fig. 1.13 Diamond-coated 'card'.

FILE CARE

Files do not last for ever, but, if they are used with respect and stored so they cannot knock against each other, they will last longer than you might expect. A good tip is to keep new files for use on brass, where sharpness really affects how they work. As the files dull, move them on to steel and iron. If they clog up, particularly after filing aluminium, they can be cleaned with a 'file card', a board with stiff brass bristles, or a stiff brass or soft iron brush. Work these along the lines of teeth to remove the clog. Another way to clean a file is to take a length of brass bar and work the end of it across the file in the same way. It should rapidly adapt to the shape of the file and start clearing out embedded material.

FILE RACK

All conscientious workers realize the need to keep files so that they do not knock against each other and become blunt. Large files can be hung on hooks, stored in stacked tubing or kept in divided drawers. Small files such as Swiss or needle files usually come in plastic wallets, but these can become brittle with age and often fill with swarf. There is a need for something a bit more accessible that also makes it easy to choose between several files of the same pattern. Many people tend to keep some files for 'best' and others for regular use, meaning that they need quick access to those particular ones.

One basic solution is to keep the files slotted into a rack made from a block of wood with holes in it, but this approach does not make it easy to see which file is which. Another solution is a 3D-printed file rack. The simple design illustrated here was created by 'extruding' the U-shaped cross-section, then subtracting a grid of rectangular blocks to create the spaces for the files. Initially, the rack was a bit crowded and the holes were

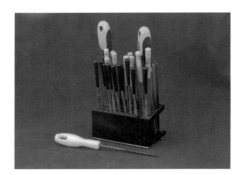

Fig. 1.17 3D-printed rack for small files.

Fig. 1.18 Specialist lathe file. It must be used with eye protection and with a handle fitted.

Fig. 1.19 Close-up of the single-cut surface of the lathe file.

a little too narrow for triangular and round files, but correcting this was easy, simply by stretching the whole design by a few per cent in its y-axis.

To minimize overhangs, print the rack on its front face. There should be no difficulty with bridging the relatively small holes for the files. Naturally, the as-printed plastic rack will be rather light, and it would be unstable when loaded with files. In this case, some nicely finished (but awful to machine) bright flats with rounded edges provided suitable material. This otherwise tends to get used for fabricating bearing pullers, brackets and the like rather than for permanent jobs. Two sections were cut to suit with the ends tidied on a finishing belt and superglued to the bottom of the rack. The result is as strong and stable a result as you might wish for!

FILING IN THE LATHE

The use of a file on rotating work in a lathe is usually to be avoided, as there is a high risk of injury if the file hits a rotating part or catches on the work. With due care, however, filing work in the lathe can be done safely – indeed, it is probably safer than using loose strips of emery cloth, a practice that results in many industrial injuries as fingers get pulled into moving machinery.

A clear view of the work and paying suitable attention are vital, and any file used with a lathe must have a secure and well-fitting handle. If a file is stuck by a moving jaw or otherwise pushed backwards, a bare tang can be driven through the palm of your hand or even into your wrist, with potentially awful consequences. A large, well-fitting handle is ideal. The very small handles that are sometimes provided with small files cannot always be trusted to give adequate protection or control of the file.

Observe obvious precautions: make sure the chuck guard is in place, and keep the file and items such as sleeves and fingers well away from the chuck. Always avoid using a file close to the chuck jaws, as contact can damage chuck, file and yourself. Eye protection is essential. When using the file, you can improve your control by gently holding the far end of the file with your left hand. Don't leave the file still at any time as this may create a soft or blunt spot on the file. Gently apply forward strokes and don't rush; let the machine do the work. For a straight chamfer, hold the file at a fixed angle; for a gentle curve, twist it sideways as you move it forwards.

Although you can potentially use any file, a proper lathe file will give better results and will be easier to control. The Tome Feteira lathe file in Fig. 1.18 shows the single, rather steep cut of a lathe file. This pattern is chosen to be much less likely to grab. It also requires constant, gentle pressure away from the chuck to keep it in place. This means that if it loses contact with the work you have a natural tendency to push it away from the chuck. With its handle, the lathe file is about 400mm (16in) long, has two 'safe' edges, and feels a little heavier than most files. I find it much easier to control than an ordinary file for even relatively delicate work.

Many people use emery tape to put a good finish on turned parts without realizing just how nasty the results of it catching on the work or chuck can be. The HSE offers some good practical advice, which in short is to use emery tape glued to a wooden backing. It can then either be used like a lathe file or levered against the work using a support. (Incidentally, the HSE's guides

Fig. 1.20 Emery tape.

to safe working practice are highly recommended. Far from being 'nanny-ish', they are full of practical advice and offer a sensible approach to safe working.)

HANDLES FOR SMALLER FILES

To reiterate, the use of file handles is not simply best practice; it is essential for safety and ease of use. Unfortunately, most handles for small files are too small and ordinary handles are too large. A better example is thing number 1866871, designed by Jangles1981 and downloaded from Thingiverse. It is one of three styles of file handle under this object (Fig. 1.21) and, like many objects on Thingiverse, it is shared under the Creative Commons–Attribution–Share Alike licence.

These handles are well designed and printed, with adequate fill. Pushed on to 3mm file, they seem to hold them firmly. The Cura settings used are shown in Fig. 1.21. The flatter designs print more quickly but are perhaps not as comfortable to use. Perhaps the most important factor to watch for is that the handle should have a suitable socket to suit the file you want to use it with. Some designs benefit from heating the tang just enough to soften the print as it is inserted, to ensure a good grip, especially those for files with a taper tang. Be careful not to overheat and soften the file.

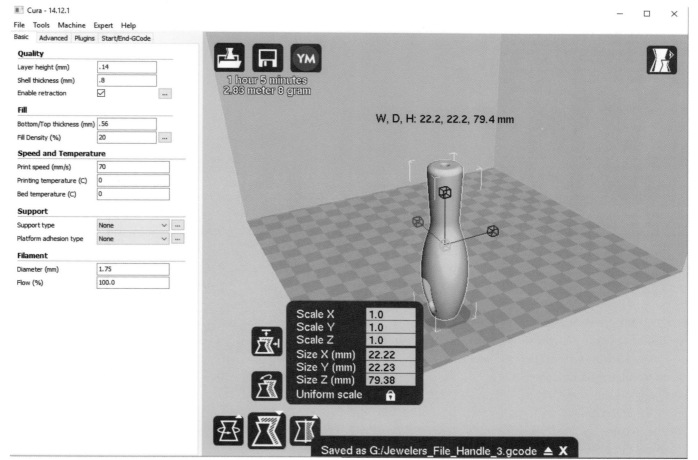

Fig. 1.21 File handle in Cura, a 3D print 'slicing' program.

THINGIVERSE

Enter a term such as 'file handle' into the Thingiverse search box and you will be rewarded with many designs; handles large and tiny for files big and small. Be aware, though, that not all objects on Thingiverse are as well designed as the example of thing number 1866871. Some of them are virtually unprintable! The best approach is to search and find things that look like they will suit your purpose, then download them and try slicing them to see if there are any issues before committing to printing them. If the listing has an 'I Made One' section, you can see if other Thingiverse users have had success with the design, and the 'Comments' section may give useful tips on the best settings for a good print.

Another issue with Thingiverse is that most objects are available only as STL files. This is fine if you just want to print an object, and perhaps scale it or mirror it, but not as useful as having the original 3D model in a CAD format.

One way this is addressed is that some objects on Thingiverse are 'parametric', which means they are written as computer programs. You enter dimensions or numbers for elements such as the number of teeth on a gear, and the program then generates an STL of the object to match your input.

Many of the objects on Thingiverse are covered by the Creative Commons–Attribution–Share Alike licence, which allows you to download and print an object, as well as modify it, 'remix' it, and share your own version, as long as the correct attribution is given.

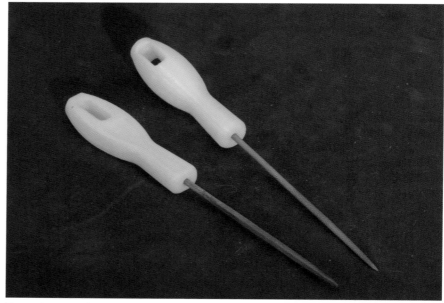

Fig. 1.22 Two styles of 3D-printed handles for small files.

FILING BUTTONS

Filing buttons are a simple yet astoundingly useful aid to shaping components by hand. They are just round discs of metal fitted either side of a hole to guide hand filing and help with the production of a neat, round end. For a rough and ready result, just using a cap screw or cheese-head screw might be sufficient (Fig. 1.23), but with just a little more care you can achieve excellent results.

Obviously, in order to get a good result, the button should be exactly the same width as the bar, and the bolt that holds buttons and bar together must be a good fit in all three parts. This is not hard to achieve, even if you are using a plain screw to hold the parts, but do take time to measure the screw first. Many screws (especially metric ones) are significantly smaller than their nominal top diameter.

Filing buttons made from silver steel, hardened but not tempered, will be glass hard, the file will skid off them without damage, and they will not need to be left free to rotate. You can use softer materials, but if you do the buttons need to be able to rotate so that the file does not wear them away; this means a sloppy fit at the expense of a little

Fig. 1.23 Improvised filing button from a cap screw.

accuracy. If you do not have material of the correct diameter, you can simply turn some down – the surface finish is not critical. Drill through a good snug fit for the securing screw, and part off two slices. These should not be too thin, like washers, instead aim for similar proportions to a full nut. Use the same drill on the component to be shaped – you can enlarge the hole to a larger size later if required.

Using the buttons is simplicity itself. Fit hardened buttons securely in place, or leave softer ones a running fit. Hold the part securely in a vice and file away, guided by the buttons. Interestingly, the buttons control the cut taken by the file, so getting a good finish is usually easy, even with a coarser file than usual.

Normally you would use a flat file, but you can use more than one button to create more complex shapes with other types of file. The balanced crank in Fig. 1.25 was made using three pairs of buttons in two sizes and the concave sections between each pair of buttons were finished with a half-round file. Fig. 1.25 shows the crank with its edges pro-filed, then treated with a sanding drum followed by a felt mop with polishing compound. The hole for fixing the buttons to the balance weight cannot be seen, as it has been plugged by a piece of similar steel turned to a driving fit. The plug is invisible after polishing.

Finally, keep your buttons paired up on their matching screw. You will soon build up a little collection of useful sizes. If properly hardened, they should last a lifetime.

Fig. 1.24 Hardened filing buttons from silver steel.

Fig. 1.25 The finished crank arm after filing and polishing.

2 Measuring and Marking Out

One of the questions many beginners ask is, 'How accurately should I be working'? Some say you should always work as accurately as you can because this is the way to develop your skills, but in practice the best policy is to consider each task in turn and work to an appropriate level of accuracy.

The term 'accuracy' refers to an assessment of how close something is to its nominal value. Compare it to 'precision', which describes to what degree the value of something is measured or a process is carried out. It is possible to be accurate without being precise, and vice versa. An example of accuracy without precision might be the making of a set of fixing studs for a model – they all need to be the same length for appearance's sake (accurate) but the exact size is relatively unimportant within a couple of millimetres (it can be imprecise). In this case, a simple measurement with a

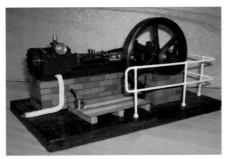

Fig. 2.1 A model steam engine with some accurate and some precision parts.

rule will be adequate. In contrast, to be a close fit in a hole, a pin will need precision to match between the parts, but the exact size of the hole and pin may be relatively unimportant.

In the past, tolerances were often implied by how sizes were specified: 'one and a quarter inches' would be a rule dimension, but '1.250in' implies greater care in measurement. Alternatively, expressions such as 'full' and 'bare' for slightly over- and undersize would be used and fits specified with terms such as 'free-running', 'sliding', 'push' and 'force'.

While such vague terms work perfectly well for one-off jobs, such as those that are often carried out in a private workshop, the demands of modern industry mean that every dimension now has a specified tolerance and/or surface finish associated with it. This is because in mass production parts need to be easily interchangeable and made to traceable standards. In your own workshop, you are generally more interested in making sure parts work with each other, so, as long as they are well-proportioned, small differences in size from the plans are rarely critical. This is reflected in most hobby designs, where it is usually left up to the maker to decide what fits are appropriate.

Exact sizes are less important than the quality of fit – it is far more important

for a piston to be a good, close sliding fit in its cylinder than for it to be exactly a certain size. Equally, a pin that is to be a force fit in a hole and a spindle that will rotate in a bush need very different standards of fit. The difference in size between two mating parts to give them the appropriate fit is often referred to as an 'allowance'.

Obviously, there are exceptions, such as when two sets of parts must interact or when you are fitting your parts to bought-in items, such as a turning shaft to fit a standard bearing.

In practice, most measurements on plans will be specified in a way that indicates how accurately you need to work. Typically, most parts needing to be an accurate size will be specified to a precision of 0.02mm or 0.001in (universally called a 'thou'), and most machine tools will have their movements graduated in steps close to these sizes. In contrast, dimensions given in fractions of an inch or to the nearest millimetre can usually be marked off using a rule – although you will need to watch out when two parts have to be matched to the same size.

This does not mean you won't sometimes see plans with parts such as a shaft and its hole specified as the same diameter, for example. In such cases, use common sense – an off-the-shelf steel bar will usually be a good but easy fit in a hole made by a standard H7 reamer (the grade that is usually supplied) of the same nominal diameter.

MEASURING RULES

It is worth having a small collection of workshop rules (not 'rulers' in engineering circles), from a cheap 'hack' used to check whether a bar is 25mm or an inch in diameter to a more accurate one that reads down to 0.5mm or 0.02in. When using a rule, always make sure you look directly down at the graduations, not at an angle, or the thickness of the rule can introduce a surprisingly significant 'parallax' error.

My own collection of rules has grown over the years, from just a few to quite a large number; Fig. 2.2 shows three different ones. My parents were somewhat taken aback when as a boy I asked for an engineer's 12-inch rule for Christmas one year. I used it largely for modelling in plastic card and balsa, and I still have it today, although it is somewhat stained. Quality rules are not expensive for what they are; beware the cheapest, which may have crude, stamped (rather than engraved or etched) graduations and irregular subdivisions. Worst of all is the apparently accurate rule whose first division is slightly out, making all absolute measurements wrong.

If resources allow, it is worth having some shorter rules for measuring in tight spaces and a thin, flexible rule for measuring and marking curved surfaces.

Everyone has a few items in their workshop without which they would be lost. For me, one of these is a tiny rule. When working on small items I was regularly frustrated by the difficulty of

Fig. 2.2 A selection of quality rules.

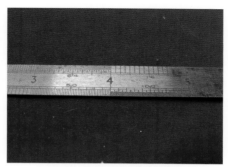

Fig. 2.3 Example of high-resolution rule gradations (0.01in).

Fig. 2.4 Flexible rule for use on curved surfaces.

getting callipers or even a 6-inch rule on to the job. I have read of much smaller rules, down to just a quarter of an inch long, but these seem to be very difficult to find.

Many workshops will have an old or cheap carpenter's square that has lost, or never had, any great degree of squareness. If you have one with a blade graduated in inches and millimetres, cut

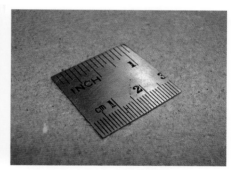

Fig. 2.5 'One-inch' rule for inaccessible spaces.

around 30mm from the end of the blade (*see* Fig. 2.5). Keep the little rule on the headstock of your lathe. It is unlikely to be finely graduated, just down to 16ths, so it will not be suitable for precision measurement. Instead, it will be invaluable for keeping an eye on the progress of a job, which might otherwise demand continually winding the saddle back and forth to make space for a larger measuring instrument. Do not agonize about damaging one of your favourite rules for this purpose. Next time you are at a boot sale, market or pound shop, buy a cheap 'disposable' combination square. Chop it into as many different length sections as you like and throw the rest into the scrap box.

CALLIPERS

Basic Callipers

Traditional callipers are a simple device used by setting them to the width of a workpiece then measuring the distance between the points using a rule. Fig. 2.6 shows a typical sprung pair, as well as an older-style pair that relies on a friction joint between its legs. For the most accurate results, the calliper should be adjusted until only the very

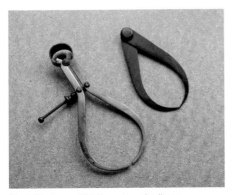
Fig. 2.6 Spring and friction external callipers.

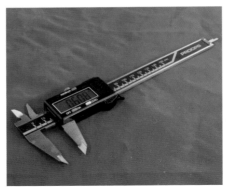
Fig. 2.8 Digital callipers.

Fig. 2.11 'Left-handed' callipers are very useful for lathe work.

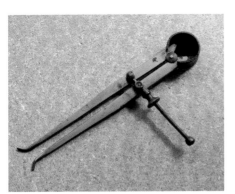
Fig. 2.7 Spring-bow internal callipers.

Fig. 2.9 Typical digital calliper readout.

A 'left-handed' pair of digital callipers is a handy item in any workshop. These make measuring work held in the lathe much easier.

MICROMETERS

Using a Micrometer

lightest drag is felt as it is moved across the work to be measured.

Internal callipers are used for holes, slots and the like.

Measuring Callipers

In most workshops the most frequently used and most useful measuring devices is a measuring calliper, which will usually read to 0.02mm or 0.001in. Old-style vernier callipers are read using special scales, but most hobbyists today use digital callipers, which can be read at a glance and easily convert between metric and imperial measurements. As well as measuring length between the main jaws, callipers also have pointed internal jaws for use inside holes (but

Fig. 2.10 Using digital callipers to measure lathe saddle movement.

beware of trusting these for small holes!), a depth rod and even the ability to measure steps using the back of the jaws. Fig. 2.10 shows a digital calliper being used to measure the distance from the tailstock to the saddle of a lathe, to help with making an accurate cut.

It may seem that a good rule and a set of quality callipers will do everything, but from time to time it is useful also to have a micrometer, often reading down to as small as 0.002mm (0.0001in), particularly when shafts and bearings have to be turned accurately. Micrometers work by gently trapping the work to be measured between two very flat anvils, one moved using a high-precision screw thread. Older micrometers can be cheap to acquire second hand but are trickier to read and may be worn; if you can find a good example, though, they are a pleasure to use. Again, digital versions are much easier to use, but remember to check that they are properly 'zeroed' before each measurement. Fig. 2.14 shows a 50mm (1–2in) micrometer that has to be zeroed using an accurately sized tungsten carbide test bar.

Fig. 2.12 Starrett mechanical micrometer.

Fig. 2.13 Digital 0–25mm micrometer.

Fig. 2.14 Digital 25–50mm micrometer.

While these tiny distances (less than a hair's breadth) sound impressive, the truth is that, unless your measuring technique is exceptional and you have a temperature-controlled workshop, you are unlikely to be able consistently to work to such levels of accuracy. Unless you have a very well-developed

and consistent touch, always use the small knob at the end of the handle when tightening the micrometer on to an object to be measured. This operates a slipping clutch at a consistent level of pressure to help prevent any discrepancies.

A Micrometer Stand

When measuring very small objects, it is easier to hold the item in some sort of stand, to offer it up to a micrometer. A makeshift approach is to use card to protect the frame of the micrometer and gently hold it in a vice, but this is not always convenient and does not present the micrometer at a convenient angle for easy reading. A better solution is a simple micrometer stand.

The stand here is made in two parts: the base and the clamp itself. The main requirement for the base is that it should be stable, so it is ideally made from a suitable offcut of steel or cast iron. It can be as plain or decorated as you wish. Recessing the underside will avoid any issues if your lathe does not face things concave, and will make it more stable when it is placed on a surface that is not perfectly clean or flat. It does not need to be huge; the base

in the example here is only 50mm (2in) diameter and works with a 25mm (1in) micrometer. A 75mm (3in) base would be even more stable and work with larger micrometers as well. The centre of the base is drilled and tapped with an M6 thread.

The clamp itself is made from a few 3D-printed parts. Although these could equally well be made from metal, 3D printing allows rapid revisions to be made to achieve a good result. The parts are:

◆ The yoke is screwed to the base with an M6 screw and allows the clamp proper to be adjusted for angle.
◆ The fixed jaw has an M6 thread printed in, so that it can be clamped to the yoke at the required angle. Printed threads are more than adequate for this sort of light duty.
◆ The moving jaw has two semi-circular cut-outs that locate on bosses on the fixed jaw, to keep everything in line, and uses a second M6 clamping screw from the scrap box.

Because of the plastic construction, it is not necessary to line the jaws to protect the micrometer from scratches. If you don't have any suitable clamp screws

Fig. 2.15 Micrometer stand allows hands-free use.

Fig. 2.16 Illustrating construction of the micrometer stand.

with knobs, just 3D-print some knobs with a counter bore for the heads of M6 cap screws. The screws can be epoxied in place.

MEASURING HOLES

One of the trickiest measuring tasks is sizing holes. Your starting point should always be a clear understanding of how accurate a hole needs to be. If simply drilling or reaming a hole is going to provide the required tolerance, it is clearly a waste of time to spend a lot of effort trying to make it to a greater accuracy. If a hole does need to be made more accurately, or you need to measure a hole, there are a number of different approaches to consider, with varying degrees of simplicity and accuracy.

For smaller holes, a very easy option is to use the shank of a twist drill – just find the drill that gives the best fit. Be aware that the shanks of many drills are slightly undersize, so it is worth checking them with a micrometer or callipers. It is obviously also possible to use metal bars, but these are rarely available in such a wide range of sizes.

Fig. 2.18 These pin punches can also be used to gauge hole sizes.

Fig. 2.20 Taper gauge measuring 25mm bore of a tube.

Another possibility is to use a pin punch, especially when deep holes need to be checked, but these are almost always rather smaller than their nominal diameter; for example, my 1/8in pin punch measures at 0.121in.

For larger diameters, a tapered hole gauge can be used. The accuracy of these is modest and does depend on the end of the hole being accurately square. However, they are easy and rapid to use.

One way to measure holes that is usually more accurate than a taper gauge is to use the 'blades' on the 'back' of a pair of callipers. To get the most accurate reading, gently rock the callipers until you can feel that their edges

have settled on the sides of the hole. Because the edges of the blades are often not completely sharp, they may not be particularly accurate on small holes, but the figures for larger holes are usually consistent with those calculated by other methods.

More accurate hole gauges can be bought, but they can also be made. It is important to have a very good surface finish and, ideally, precision-ground mild steel or silver steel should be used. It is a good idea to chamfer the end, leave a narrow (1–2mm) strip at full diameter and then machine a shallow groove just beyond the strip (*see* Fig. 2.21). This makes it much easier to insert and align the gauge with a hole when it is a very close fit, with much less chance of jamming.

Telescoping bore gauges are T-shaped tools with two sprung probes on a handle. They are inserted into a hole and gently rocked until the ends lie across the hole diameter and then locked by turning a knob at the end of the handle. A conventional micrometer or calliper can then be used to take a reading across the ends of the gauge. Both getting the original setting and taking a reading are quite tricky actions to do well, although

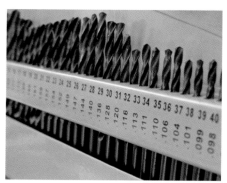
Fig. 2.17 Number drills can be used to gauge the size of small holes.

Fig. 2.19 Taper gauges provide a reasonably accurate way to measure larger holes.

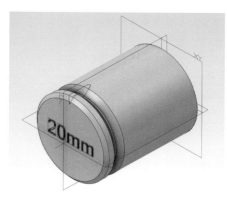

Fig. 2.21 Accurate bore gauges can be used on a cut-and-try basis to size holes, or to 'prove' the size of pre-machined holes.

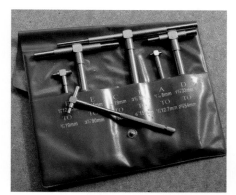

Fig. 2.22 Telescoping bore gauges can be used to measure larger holes in company with a micrometer or digital calliper.

the convex ends of the probes help. It is worth practising until you can get consistent readings from a selection of holes. A useful tip is to set the gauge slightly wide and the clamping knob so that the probes are only just held in position. The end can then be gently rocked sideways in the hole, pushing the probe ends into position. It is very important not to overtighten the clamp when doing this.

The 'cheap and cheerful' alternative to telescoping gauges are the old-fashioned internal callipers. These are also used by being set slightly larger than

the hole diameter and then rocked to settle them at an accurate diameter, which can then be checked with a micrometer, or even against a rule if the measurement is not critical.

Bore micrometers are similar to telescoping bore gauges in their method of use, but actually incorporate a small micrometer barrel to which additional extension pieces can be added, to cover a range of bore sizes in small steps. Unlike sprung probes, a micrometer bore gauge has to be used on an adjust and try basis. This can make measuring an existing hole a rather drawn-out process and they are better suited to checking progress in opening up a hole to a diameter already set on the gauge.

An intermediate alternative to the telescoping probes and the bore micrometer is a device with an integral dial

indicator and two probes. The probes are set to a reference diameter within a range of around 15mm, and the dial then indicates a size difference of around plus or minus 1mm either side.

DIAL GAUGES

Dial gauges have a plunger and a dial test indicator (DTI), which operates with a small ball-ended lever. They are particularly useful for tasks such as ensuring work is truly concentric with the lathe spindle. In use, they are usually held on a stand with a heavy or magnetic base. It is also possible to use them in other ways such as held in the toolpost of a lathe or the spindle of a milling machine – an arrangement which is very useful for aligning a vice on the mill table.

Fig. 2.23 Micrometer bore gauge.

Fig. 2.24 Head of the gauge showing extension piece.

Fig. 2.25 A pair of dial gauges.

Fig. 2.26 Dial test indicator (DTI).

Fig. 2.27 Using a dial gauge to measure runout.

Dial gauges can be used to make absolute measurements of distance, but, except for some specialist types, dial test indicators (DTIs) are generally for comparing measurements such as the height of two components.

HEIGHT GAUGES

There are two types of height gauge: simple height-adjustable points and the measuring type. The latter are essentially similar to callipers fitted to a solid base; the digital versions are particularly useful as they can be set to zero

Fig. 2.28 Digital height gauge.

at any height. Simple points are set to a height by comparison against a rule held vertical against a stand or just a square block.

SURFACE PLATES

A proper surface plate is an accurately finished flat metal or granite surface, used as a base with height gauges and other measuring equipment for tasks such as marking out parts and castings for machining, or checking the size of parts. While professional surface plates are incredibly flat, for general workshop use many people get away with using a slab of granite kitchen worktop or a piece of plate glass. Another option is the very flat piece of glass from an old computer scanner, fitted in an accurate wooden frame and supported by a thin layer of felt to stop it distorting.

Fig. 2.29 A variety of measuring tools on a surface plate made from a piece of polished granite worktop.

MARKING OUT

If you are making items to any sort of plan, from a scribble on the back of an envelope to a computer model, you will find that you need to accurately mark some parts as a guide to machining. The essential tool for this is the 'scriber', a pen-like tool with a sharp, hardened

steel or carbide point. Some measuring tools such as height gauges have a built-in point. When using a rule, watch out for parallax errors (*see* above). If you run the scriber along a rule to mark a line, hold it at 45 degrees so that it goes right into the corner between the rule and the work.

Hot rolled steel sheet has a dark colour that shows scribed lines well, but they will be less easy to see on bright steel. To increase the visibility of scribed lines, it helps to colour the surface to be marked out. On steel and aluminium alloy, a product called 'marking blue' can be sprayed on or applied by brush or cloth, to ensure that lines are bright and clear. A black or blue spirit marker serves a similar purpose. Cast iron does not give bright scribed lines, so many people use white spray paint or even correction fluid on this material to prepare for marking out.

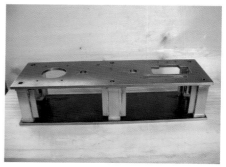

Fig. 2.30 This model steam engine entablature required accurate marking out.

Fig. 2.31 Carbide- and steel-tipped scribers.

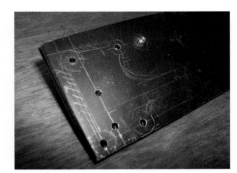

Fig. 2.32 Marking out on hot-rolled steel.

Fig. 2.33 Marking blue improves the visibility of lines on bright steel.

This process of marking out is greatly helped when the plan takes as many measurements as possible from a 'reference point' or face. You can use a rule or calliper to scribe lines at known distances from the end of a bar, for example, or use a height gauge and a surface plate to scribe lines on castings or even metal plate held at right-angles to the plate. With castings it often pays to machine the reference face(s) flat, so the casting sits solidly without rocking.

A RULE STAND

A valuable companion for the surface gauge is a stand to hold a rule vertical and facilitate setting the gauge to an accurate height. Again, there is a simple solution: a nice, square piece of metal and a clamp to hold a rule in place. The

stand in Fig. 2.34 – based on one seen in an old issue of *Model Engineer* magazine – has an integral clamp that makes it much easier to use.

The clamping piece is banjo-shaped, with the round end running in a similarly shaped groove (Fig. 2.35). It is moved by a simple screw thread operated using the knurled knob. An option collar on the screw part keeps it in place when there is no rule fitted. Both the clamp piece and the block have shallow dovetails, to ensure the rule is pulled back snugly against the block. Machining these is an ideal task for a small shop-made dovetail cutter, offering a good way to get some practice in cutter making, as the shallowness makes the job relatively undemanding.

None of the dimensions is critical: a block about 25mm (1in) square is a

Fig. 2.34 Rule stand.

Fig. 2.35 The moving clamp of the rule stand.

good size for 150mm or 6in rules, while longer rules would require something larger for stability. Otherwise, just try to make the clamp a close fit in the body to give a smooth action.

PUNCHES

Centre punches are used to make relatively large indentations in metal, typically to provide a depression to guide the start of a drill. The end of a centre punch is usually ground at about 90 degrees to give a relatively large depression without requiring the end to penetrate far. This also makes the tool quite robust.

Automatic centre punches are useful tools, as they can be operated with a single hand. They contain an internal spring so that, when pressed down on to a workpiece, the punch delivers a firm but consistent blow to create an indentation without the use of a hammer.

A punch with a more tapered tip – about 60 degrees – is often called a dot punch. This makes a smaller depression and is useful when marking out, to provide a guide for filing to a line. The idea is then to file until you have 'halved' the dots.

Fig. 2.36 Various types of punch (L to R): prick punch, dot punch and centre punch.

A punch with a very sharp end, say 30 degrees, is known as a prick punch and is used for fine marking out. It can, for example, provide a guide for one end of a pair of dividers when marking an arc. If required, the impressions made with a prick punch can be enlarged with a dot or centre punch. Prick punches are almost as sharp as scribers, and therefore are quite fragile compared to larger punches.

A light tap with a small hammer is sufficient to mark a point. For best results, follow up with a heavier, blunter centre punch to create an impression that will guide the point of a drill. Typically, this will give an accuracy of about 0.1mm (0.005in), with care.

A magnifying glass or close-work binoviewers are a great help when doing fine marking out. It is even possible to get hold of small magnifying lenses that clip to a centre punch. In practice, most people find that a pair of very strong reading glasses are a comfortable solution.

A handy tip when wanting to punch a mark at the intersection of two scribed lines is to run a prick punch down one line until you feel it 'click' into the crossing point. This is surprisingly easy to judge, even with very lightly scribed lines.

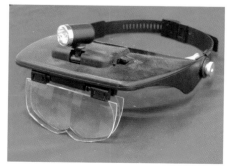

Fig. 2.37 Visor for close work.

Making Punches

Punches can be made from carbon steels and hardened and tempered in the same way as cutting tools. A 'deep straw' temper is appropriate for most cutting tools and can be used for scribers. However, this is still too brittle for striking tools such as punches and chisels, which are best tempered to purple or dark blue.

Generally, these tools benefit from being hardened, tempering by heating, then cleaning, then heating the shank gently and letting the oxidation colours develop and 'run down' towards the tip before quenching. This means the working tip will be the hardest part of the tool, while the shank will be softer but more resilient and less likely to shatter.

Turning the end of a long thin punch can be tricky if it is extended a long way from the chuck. It is often best to put the angle end on the punch with the work only protruding slightly, and then extend the work to shape the body. Alternatively, the punch blank can be left blunt and then ground to a point after hardening, obviously taking care not to overheat it.

Punches can be made from material of any regular cross-section: a square body is less likely to roll away, while a round one will feel more comfortable in the hand. Hexagonal stock is often a good compromise if you have it in the right size. If you want to ensure a good grip you can turn grooves around the body or knurl all or part of it. Do not forget to round the edges of the upper part so that there is no danger of raising a burr when striking the punch.

Bell Punch

A bell punch is a centre punch within a conical housing. It is used for marking the approximate centre of circular work, by simply holding the body down on to the end of the bar and tapping the punch. These are not intended to accurately determine the centre but are very useful as a practical aid to setting up larger round work in the lathe. With a little care, they can be used on other regular shapes such as squares and hexagons.

Fig. 2.38 Bell punch.

Making a bell punch is straightforward, and a good use for a large bar end. All that is required is a well-fitting hole for a simple smooth-sided punch that is concentric with a deep conical recess. Something of between 50 and 75mm (2 and 3in) is most useful, but the relatively large amount of turning required to produce the bell makes working on a larger one a long job, especially if you reduce the outside of the body as well. Castings are sometimes available to make the work easier, but these often cost as much as buying a completed punch.

Letter and Number Punches

Not very practical to make, unless you are very skilled, but useful to have, are letter and number punches. Using these to produce a neat result can be very difficult, as they are almost impossible to place accurately without some sort of guide. A good firm tap is required to ensure the character is marked in a single action. Note that smaller characters, especially '1' and ",, need less force than the rest of the letters and numbers.

The simplest type of guide is a bar with a groove milled across it to match the body size of the punch. This can be held in a lathe's toolpost, for example, to allow marking on round surfaces, as in Fig. 2.40. For flat surfaces, it can be a

Fig. 2.40 Dial marked out using a jig to guide number punches.

little more challenging, but it is usually possible to set up a jig using metal blocks. The biggest challenge is getting good spacing between the characters; this normally requires a degree of experimentation.

SURFACE GAUGES

A Simple Surface Gauge

A surface gauge is an essential aid to accurate marking out. The minimum needed is a stable, flat-bottomed base with a pillar and clamp to hold a scriber, allowing lines to be drawn truly parallel with the top of a surface plate. Such a device has many other uses, such as helping set work in a four-jaw chuck or any other task where the ability to 'mark' a point in space with the end of a scriber is needed.

The best material for the base is steel or cast iron – it needs to be heavy for stability, and hard enough not to be easily damaged or worn. It does not need to be fully hardened. If it is cast

Fig. 2.39 Number and letter punches.

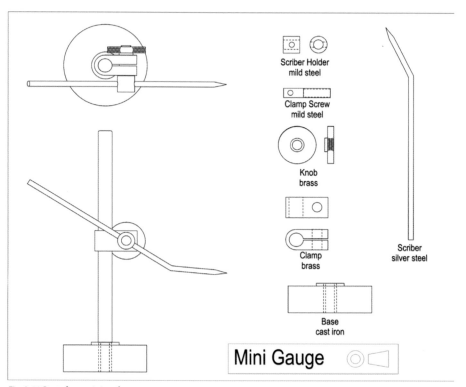

Fig. 2.41 Parts for a mini surface gauge.

iron or mild steel, you can take a tiny skim off the bottom to true it up after 20 years or so. The hole for the pillar is best bored in the lathe. In truth, because a surface gauge has inherent accuracy, the angle of the pillar does not matter a jot, but it is not worth spoiling the ship for a ha'porth of tar. In other words, even non-critical tasks should be done to the best of your ability, so that when a critical job does come along you will be well prepared. Try at least to make this pillar vertical!

Silver steel of 1/4in (6mm) diameter is fine for the pillar. Do not harden it, as this may cause distortion. Precision-ground mild steel would do just as well. Thread both hole and pillar in the lathe – 1/4in BSW, 0BA or M6 will do fine. Carefully open out the upper part of the hole, so that some of the full diameter of the pillar enters the hole, holding it in line. You can also use a force fit, a retaining adhesive or even a set screw to fix the pillar in place. Whatever you decide,

produce the hole in the lathe, not on the drilling machine.

If you have doubts, hold the assembled pillar and base in the three-jaw chuck and take a light skim to make sure it is both true and flat. Make sure it does not wobble when placed on a flat surface, such as the bed of your lathe.

The simplest and best clamp for such a device is shown in the drawings. A central shaft is threaded at one end and has a hole for the scriber at the other. This is threaded through the body of the clamp, a sleeve fitted around the scriber end and the scriber inserted through both sleeve and shaft. Provided the clamp body is a good fit on the pillar, then a simple knurled nut on the thread will pull the parts together with ample force to secure everything in place. Brass is a good material for the clamp body (it can be finished by draw filing), while the other parts of the clamp can be mild steel.

The scriber should be made from silver steel (drill rod in the USA) – 2.5mm (3/32in) is a good size. Cut it to length and carefully turn a taper on each end, finishing with a tiny 60-degree point. If you use files to do this, take great care. Make sure they have secure handles and that you wear safety goggles – if the file makes contact with the chuck

jaws, it can be thrown into your face or pushed into your hand. Bend the scriber about 30 degrees near one end, to help when setting it low down. Hardening such a long, thin piece can be difficult. It may be easier to heat it to a bright red along its length using a gas cooker or a camping stove, rather than a blowtorch, picking it up at the centre with a pair of pliers and plunging it rapidly into a bowl of water. An alternative is to heat and quench each end separately, but this will leave the centre of the scriber soft, and it may bend in use.

Similarly, there are two approaches to annealing the scriber. Heating the centre until the ends turn a dark straw colour will leave the ends hard and the centre less so, but again the risk is that the centre will be easily bent. Scribers may be cooked at gas mark 6–7 (about 200 to 220°C), but you need to remember than an oven is not a precision instrument! Try putting the work in at a lower setting, and then 'tweaking' it up every ten minutes until you get the right 'bronzy' dark straw colour, which will indicate an even temper along the whole length.

A Micro-Adjustable Surface Gauge

The basic 'scriber on a post' works well, but sometimes a more robust device is needed, and often the ability to make a fine adjustment to the height of the scriber is useful. The answer lies in a design that has been made (with many detail variations) by hundreds of apprentices over the years (Figs 2.44 and 2.45). It relies on the fact that the pillar does not need to be vertical. As long as the base of the body is flat, and there is no play

Fig. 2.42 The mini surface gauge.

Fig. 2.43 Detail of the clamp and scriber.

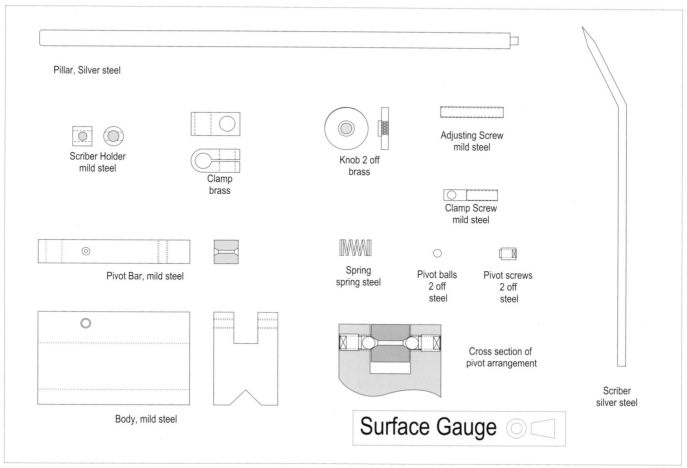

Pillar, Silver steel

Scriber Holder
mild steel

Clamp
brass

Knob 2 off
brass

Adjusting Screw
mild steel

Clamp Screw
mild steel

Pivot Bar, mild steel

Spring
spring steel

Pivot balls
2 off
steel

Pivot screws
2 off
steel

Cross section of
pivot arrangement

Body, mild steel

Scriber
silver steel

Surface Gauge

Fig. 2.44 Parts for a micro-adjustable surface gauge.

anywhere, the whole thing is inherently accurate. At least, this will be the case in Euclidean space – if your workshop is on the edge of a black hole, you may have other worries! The design in the drawings uses a tilting bar in a heavy base. No dimensions are given, as nothing matters other than making the parts to suit each other, and to suit what offcuts you have and what size of gauge you need.

Both ends of the pivot hole need to be countersunk, using a centre drill of the same diameter as the ball bearings for the pivots. Make a countersunk hole to almost the full diameter of the drill, but

Fig. 2.45 The micro-adjustable surface gauge.

Fig. 2.46 Tilting mechanism.

do not drill past the tapered section of the drill. Follow up with a drill the same size as the tip of the centre drill. Now turn over the bar and carefully align the hole with the drill before clamping the bar in place. Now swap back to the centre drill and create another countersink.

The base is a section of mild-steel bar but could be a good aluminium alloy, skimmed on all sides (on a lathe or mill) to get it square. The V-groove in the base can be milled with a suitable cutter, holding the base in a pair of V-blocks. Accuracy is not essential to the function of the gauge, but try to do a neat, symmetrical job as practice for when accuracy will matter. This allows the base to sit on a curved surface (or the inverted-V bed of the lathe); if all you want is to ensure the block sits evenly on a flat surface, simply recessing the base is sufficient.

The groove on top of the base should be milled out to be a sliding fit for the tilting bar but the cross-hole for the pivots should be drilled first, as it is essential that these are in line for smooth operation. The pivot holes need to be a close fit for the ball-bearing pivots. For the record, I used two 1/8in ball bearings and tapped the holes M5 for some flat-tipped grub screws. It will now be possible to mount the tilting bar between two ball bearings held in the tapped holes using a pair of set screws. The bar should move freely, even when the bearings are held quite firmly.

Locate the correct position for a stepped hole to take the return spring under the pillar. The return spring should be reasonably strong, but it does not need to be excessively stiff. In use, the action of scribing pushes back against the adjusting screw and assists the spring. Make a suitably decorative pair of knurled nuts, one for the clamp and one to secure to the end of the adjusting screw for the tilting bar. Take a little time to set the bearings just right, and you will be rewarded with a smooth action to equal that of any shop-bought piece of equipment.

The clamp and scriber for the micro-adjustable surface gauge are identical to those for the simple gauge described above, and the scriber should also be hardened in the same way. The finished gauge is shown in Fig. 2.48, being used with the rule stand described earlier.

Fig. 2.47 Clamp detail.

Fig. 2.48 Setting the gauge against a rule.

3 Aids to Accurate Working

SIMPLE CENTRE GAUGES

The commonest items in the scrap box are offcuts of metal bar that are too small for most uses but too big to throw away. This simple but handy gauge needs no more than an inch or so of round bar. It can be made from almost any cylindrical piece of scrap that is greater in diameter than the holes in which it will fit.

It is often necessary to position the centre of a rotary table centrally on a mill or lathe. When reasonable accuracy rather than absolute precision is required, a convenient solution is a simple centre gauge. Setting up is as simple as popping the gauge into place and matching it to a plain centre in the mandrel.

As the device is easily made double-ended, one end can be made to fit the table and the other to fit your most likely accessory. In this case, one shoulder is

Fig. 3.1 Centre gauge.

a gentle push fit in the central hole of the rotary table and the other fits in the faceplate. The gauge may also be held in the jaws of a chuck.

If your rotary table has a tapered bore, start by turning a short stub taper. The locating pins should be turned to a 60-degree internal angle; this makes it easier to match them to a plain centre. They should not be too small, and if you place the pin inside a shallow recess, it will have a little protection. If you turn the gauge by holding the work in a three-jaw chuck, the accuracy of the chuck will be the limiting factor (along with your eyesight).

In use, the gauge is simply fitted to the rotary table and lined up with a dead centre or a centralized wiggler (or even a 'sticky pin' – a pin attached to the end of the mill or lathe spindle by a blob of Blu-tac or similar). While gently rotating, carefully pushing against it with the edge of a ruler or a metal bar will get it to run perfectly true. You should be able to line the end of the pin up the gauge to an accuracy of a few thousandths of an inch, although good light and a magnifying lens may be needed by those who are more long-sighted than myopic!

A handy little 'miniature centre' with a similar purpose can be made by chucking a short piece of silver steel and turning a 60-degree cone on the end

(Fig. 3.3). Centre pop it adjacent to jaw number 1 and then harden and temper it. This little slave centre will not have the accuracy or rigidity of a proper Morse taper centre, but it will come in handy for all sorts of alignment tasks. You can even put it in a drill chuck and push it into a centre pop to hold a workpiece in accurate alignment before clamping it in place. For added flexibility, drill a deep, large female centre in the other end before hardening.

Fig. 3.2 Use of a 'sticky pin'.

Fig. 3.3 Temporary centre.

Fig. 3.4 Hollow reverse of temporary centre.

LATHE TOOL SETTING GAUGES

For the best performance, lathe tools should normally be set at, or very slightly below, centre height. The miniature centre can be used as a tool for setting lathe tools to centre height by a simple visual check, and for scribing across work at centre height. If you wish to have a more positive device for setting tool heights, there are two simple alternatives.

Half Gauge

This is very similar to a D-bit, just being a length of steel bar with its end milled or filed away to half thickness. When held in a chuck in the lathe it can be rotated so the flat face is up or down as required, with the tool's tip set to just graze the surface of the gauge. Being more robust than a point, hardening is less important but will prolong the life of the gauge.

Obviously, the accuracy of such a tool setting gauge depends as much on the accuracy of the chuck used to hold it as on the level of accuracy with which the gauge itself can be made. If your chuck is particularly suspect or you want greater accuracy, you might want to consider the next alternative.

Tool Height-Setting Gauge

This is essentially a modified version of the simple surface gauge (*see* page 27). Instead of the scriber, two bars mounted face to face are used to create a double-ended setting gauge, one end facing up and the other down.

It is important that the cross piece is truly horizontal, so make the bars quite thick and ream the hole through them, to achieve a close fit on the pillar of the surface gauge.

The two bars can be bonded or screwed together, in which case only one locking grub screw is needed.

Check that when fitted both ends of the cross piece are level. It can then be adjusted for use standing on either the bed or the cross slide of the lathe, whichever is most convenient to you.

DEPTH GAUGE

Commercial depth gauges are usually either in the form of a combination device with a protractor, or as a more accurate measuring device such as the digital version in Fig. 3.7. It is also possible to use the probe on a digital calliper. However, all these devices have a fairly bulky body that can make them rather clumsy in use. A more compact depth gauge that can reliably hold its setting is an invaluable asset. It is a simple tool used to test the depth of holes or transfer the size of simple measurements from one object to another.

The construction of a depth gauge is straightforward (Fig. 3.9). The dimensions

Fig. 3.6 Two Moore and Wright depth gauge/protractors.

Fig. 3.7 Digital depth gauge.

Fig. 3.5 Arrangement for a lathe tool height-setting gauge using the mini gauge base.

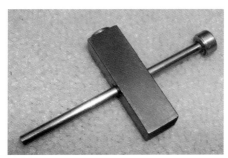

Fig. 3.8 Mechanical depth gauge.

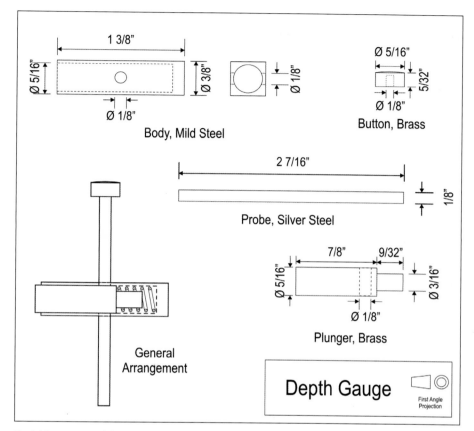

Fig. 3.9 Construction of the depth gauge.

OIL BLACKING AND BLUING

Two easy ways of putting a nice finish on the body of a tool are oil blacking and bluing. Bluing involves using fine abrasive paper to bring the surface to a good finish and heating it slowly with a small torch. The knack is to stop heating at just the right moment. Some steels do not seem to blue as well as others and for these oil blacking will achieve a more robust finish. For a small component, heat it to a dull red heat and then drop it into a tin can half-full of old cooking oil. Although doing this is unlikely to ignite the oil, it does make a lot of smoke, so please do work out of doors when oil blacking, and have a lid ready to drop over the can in case the fumes do catch alight.

can be adjusted to suit whatever is at hand, but one or two operations do demand a modicum of care. The hole for the rod needs to be truly perpendicular, and to go through the centre of the body and the plunger. Accurate marking out should be sufficient for this job. Start by drilling the body, then pop in a short piece of brass bar, followed by the plunger, and clamp it firmly in the body. Now drill the central hole through both parts, either in the lathe or a mill, for maximum accuracy. Drill the hole to a nominal size – the slight oversize of a normal twist drill is just sufficient to give a good sliding, slop-free fit for the rod in the plunger.

The second job that requires care is to finish the facing of the end of the plunger very neatly, so it is a true, flat face. The brass button on the other end of the plunger may be held on by a retainer such as Loctite. If you would

Fig. 3.10 Parts of the depth gauge.

prefer pressing the plunger to give a greater degree of freedom for the rod to move, enlarge the hole in the plunger by 0.5mm (1/64in). The various parts are shown in Fig 3.10.

CENTRE FINDERS

Centre finders are a range of several simple devices or templates for locating the centre of circular work. Most types work by allowing the scribing of accurate diameters across the work, but the bell punch works by centring round bar in a conical cavity then using an integral punch to mark the centre of the bar.

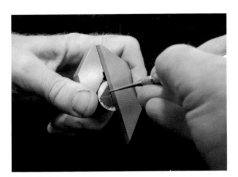

Fig. 3.11 Using a centre square.

The unusual 'square' in Fig. 3.11 is a commercial item and particularly convenient for finding the centre of small bars. Use it to scribe two diameters across the bar; where they cross should be the centre. Scribing three lines at various angles creates a small triangle around the centre of the circle if there is any error in the square. This allows the true centre to be 'eyeballed'. This three-line, rather than two-line approach is wise whenever using a new centre square, as a check of accuracy.

The variant of a centre square shown in Fig. 3.12 is actually 3D-printed as an 'emergency' tool for finding the centre of a large circular object. It is reasonably accurate in use, as it takes advantage of the high geometric accuracy of a properly set-up 3D printer. The 3D-printed version just has to be drawn with

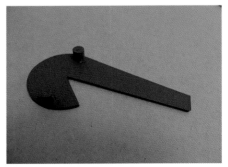

Fig. 3.12 3D-printed centre square.

reasonable care in a 3D print program so the two pins are symmetrically on either side of the centre line. The design also has the benefit of being highly scalable, so if you suddenly discover the need for a smaller or larger version it is the work of moments to make the required changes. Note that the centre edge is bevelled from below, to avoid the need to print it using a raft.

A more sophisticated version of this tool can be made by carefully marking out the overall shape on gauge plate, and press fitting two dowels or pins. The centre line should be lightly marked, but do not cut right down to the line at first. Undertake the 'three-line test' on two bars of widely different diameter and gradually machine or file the blade down until it gives accurate results at both diameters. Once this is done, it should be accurate at all sizes in its range.

If you have no centre square and are in a hurry, use a surface gauge set to approximately half the bar's diameter and use it to scribe several lines, as described above.

ENGINEERS' SQUARES

The engineer's square is an essential tool for accurate marking out and it is worth investing in quality items, like the set of three shown in Fig. 3.13. There are a lot of very cheap squares available, many of which have poor accuracy; this is especially true of cheap 'combination square' sets. A simple test of the accuracy of a square is to use it to scribe a line at right-angles to the edge of a plate, then flip it over and see if the blade is still aligned with the scribed mark.

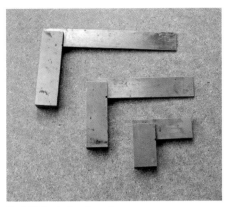

Fig. 3.13 A set of engineers' squares.

EDGE FINDERS

Despite the similar name, edge finders are used to accurately position the sides of workpieces and work very differently. They are usually used fitted in the chuck or taper of a drilling or milling machine, with a precisely dimensioned cylinder or ball at the tip of a probe. The machine is run slowly, so that the probe tip of the finder initially rotates off centre, but, when the tip contacts a workpiece, it gradually gets pushed so that it rotates truly. Once the tip is pushed even slightly past the dead true point, it runs along the side of the work. The centre of the mandrel is now half the tip's diameter from the edge of the work, typically to within less than 0.02mm (0.001in).

An electronic edge finder fits in the mandrel of a milling machine, and has a precisely centred and dimensioned sprung ball at its tip. It is not rotated; instead, when the ball contacts a workpiece, it completes a circuit and lights a lamp or makes a sound. The centre of the mandrel will then be half the ball's diameter from the edge of the work.

SURFACE PLATE

For laying out, setting up or making precise measurements, it is invaluable to have a precision plane surface that can be used for support and as a reference. It is possible to acquire cast-iron and granite surface plates that are made to high standards of accuracy. The original method of production is elegant in its simplicity – three iron plates would be hand-scraped to each other. If each combination of three plates was satisfactory, then, by definition, all three were flat.

For most model engineering purposes, a piece of thick tin-float glass, suitably supported, may be used with a high degree of confidence. For modest sizes, sections of polished granite (real stone or composite) worktop, gravestone or chopping board can be used. Some have reported that polished stone tiles also serve the purpose well. When thinking about using such alternatives, it is best to check them for flatness by sighting beneath an accurate rule placed at various angles across the surface.

Fig. 3.14 *Using float glass as a surface plate for marking out.*

GAUGE BLOCKS

Gauge blocks, also known as Johansson gauges or Jo blocks, are typically sold as comprehensive sets of precision-ground and hardened metal blocks of great accuracy. They may be used individually, or combined in stacks, as comparators for the measurement of precise distances. The surface finish of gauge blocks is such that they will normally wring together and remain in place by molecular adhesion. Typically, a set will include a pair of end blocks that should be used at the ends of a stack; these take most of the wear of ordinary use and can be replaced relatively cheaply.

There are several grades of gauge block sets, but even the most basic 'workshop grade' should be more than adequate for most hobby purposes. Quality sets are expensive, but second-hand sets that are worn or missing a few sizes can still be very useful. For some purposes, old bearing races can be used as an alternative.

GAUGE BLUE

Gauge blue is a non-drying, intense blue paste capable of being spread very

Fig. 3.15 *Old bearing races are precision-ground for parallelism and thickness.*

Fig. 3.16 *Gauge blue.*

thinly. When one surface is blued in this way and rubbed gently on a matching surface, the blue colour will transfer only at the points of contact between them. In this way, excess material can be removed. Gauge blue is not to be confused with marking blue.

TOOLMAKER'S BUTTONS

Toolmaker's buttons are a handy aid to setting up work for machining that can be used in many ways. They are accurately sized, hardened-steel cylinders, with a central hole for a fixing screw, used for laying out centres and locating parts.

For example, if several holes need to be bored in a plate in precise relation to each other, tapped holes at the approximate positions of each hole are made and used to fit buttons in position. The central bore of the buttons is oversize for the fixing screws, allowing them to be carefully moved into an exact position using gauge blocks, a micrometer, or any other means of accurate measurement. Each button is then 'brought to truth', to accurately centre the work on the faceplate of a lathe or the bed of a jig boring machine, and milled. The button is then removed, and the final hole bored out for accuracy.

BRINGING TO TRUTH

Bringing to truth is the act of setting a workpiece so accurately into alignment that further work may be carried out in alignment with reference surfaces or already machined or marked surfaces. For example, a cylinder may be brought to truth in a four-jaw chuck with a dial test indicator (DTI) running against its bore, adjusting the chuck until the gauge shows no movement when the chuck is rotated.

Another use for the buttons is to create temporary jigs on a faceplate or sub-table. One example might be using three buttons to create a 'pocket' in which workpieces can be snugged up and clamped in position under a mill or drill, allowing a run of parts to be accurately machined to the same specification.

To make a set of buttons, simply cut a number of short cylinders from silver steel and face the ends cleanly. Length is not critical – in fact, it can be helpful to have each button a different height for flexibility. Drill all the cylinders oversize for fixing screws – so to use 6mm cap screws you might drill then 8mm clear – allowing +/-1mm of adjustment in their position. Now bore or drill out the ends of the cylinders to create a recess at each end, leaving a relatively narrow lip, say 2mm wide. The recesses ensure that the buttons sit level. It can be helpful to make them deep enough to allow the fixings (typically a cap screw with a washer) to sit below the top of the buttons.

For durability, harden and temper the buttons following the usual process, to ensure that they retain their accuracy in use.

A LATHE TEST BAR

For setting up a lathe to turn accurately, it is useful to have a precision test bar that fits in the headstock. In this case (Fig. 3.17), the bar has an MT3 taper, so it is compatible with most 'hobby-size' bench lathes. You may need to use a test bar with an adaptor. If this is the case, the adaptor should be of good quality to minimize the risk of introducing an additional source of error. Their purpose is chiefly to ensure the headstock is properly aligned and any twist taken out of the lathe bed. There are alternative ways to test this alignment; anyone needing to do this is recommended to search online for 'Rollie's Dad's Method', which is more than sufficient for hobby-bench lathes, if done carefully.

Offsetting the tailstock of a lathe makes it possible to turn modest tapers, but by the same token, if the tailstock is out of line, work intended to be cylindrical will end up tapered! Accurate setting of the tailstock is also essential

for good results when drilling, reaming or tapping with the tool held in the tailstock chuck. For good results you need to be sure that the tailstock is aligned with the headstock, and that the barrel moves parallel with the lathe bed.

A lathe tailstock is typically held in place by either a long lever at the back or a large bolt through its base. Most machines allow the tailstock to be adjusted so that it is accurately aligned with the headstock spindle.

If you just need a 'quick and dirty' alignment, you can align centres fitted in the head and tailstock by eye (Fig. 3.18). If you are short-sighted or have a good hand lens or loupe, this can be a practical approach for non-critical work.

Wind the saddle up close to the headstock. With a centre in the headstock and a centre in the tailstock, their points should meet perfectly when the tailstock is slid up the bed. If you have no headstock centre, chuck a short length of round brass or steel in the three-jaw chuck and turn a 60-degree point on the end, as for the centre shown in Fig. 3.3. Once you have produced a point, mark the position of the chuck's number 1 jaw, so you can use it again. Another alternative is to use

Fig. 3.17 Precision lathe test bar.

Fig. 3.18 'Quick and dirty' tailstock alignment.

a wiggler needle, or a sticky pin set to run true, but many people find it easier to match two similar points against each other.

Another way of testing tailstock alignment is to use a thick bar with an accurate hole bored in it set between two centres. Sighting across the length of bar to align it to the edge of the top slide will give a good result. The hole must be a little less than half the bar thickness in diameter, otherwise the points of the two centres will clash.

Don't be surprised if the alignment is rather less than perfect, particularly when tried with the barrel locked in both the fully retracted and fully extended positions. You will probably find that some judicious adjustment of the tailstock is in order. Note that the vertical alignment should be spot on, as this can be adjusted only by regrinding the head or tailstock! On the other hand, you may find the side-to-side adjustment to be well out, especially on lathes where the tailstock is constructed

Fig. 3.19 Drilled bar method gives a step up in accuracy.

in two parts to allow a deliberate offset for taper turning.

If you have no dial test indicator (DTI), then setting up is a matter of twiddling until the two centres are properly aligned with the tailstock both in and out.

If you do have a DTI, lock the tailstock barrel in the extended position (retracted by a turn or so – if wound right to the end, it may twist slightly). Set the tip of the DTI on the side of the barrel and slide the tailstock back and forth by hand. You should find that you can get the tips of the two centres lined up and the barrel parallel to less than 0.02mm (0.001in) along its length.

If you want to achieve the very best results, you can turn a test bar between centres with a raised collar at each end. If one collar is 'skimmed' and the bar reversed, the other collar can be skimmed at the same setting. Both collars should now measure at the same diameter; those in Fig. 3.20 differ by 0.0003in, or about 0.001mm. With this bar between centres, you can use a DTI in the toolpost to measure the location of the collars. Once the DTI gives the same reading on each collar, that means that the tailstock is correctly set. Apply a little further 'fine adjustment' until they each give the same reading. In this way you should be able to set the tailstock so that the lathe will turn parallel to 0.01mm (0.0005in) in 100mm (4in), or better. Keep the bar safe for future use.

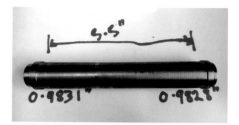

Fig. 3.20 Between-centres alignment bar.

Fig. 3.21 Using the between-centres bar to set the tailstock of a lathe.

To make a final check of the alignment of the tailstock, mount a length of brass or steel bar between centres. Take a light cut along the length of the bar. If the diameters of the two ends are no more than about 0.02mm (0.001in) different, or you cannot feel any difference when gauging each end with a pair of callipers, then the set-up is pretty good.

Finally, remember that the real test of how a machine is set up is the quality of work it produces. In modern industrial settings, the arbiter of whether or not a machine is to specification is its ability to produce work within its specified tolerance, not the measurements of its alignment and motions.

4 Heat-Treating Metals

Before you embark on making your own cutting tools, you need to have an understanding of the processes of hardening and heat-treating steel, with the emphasis on high-carbon steel for toolmaking.

'Tool steel' is a general term for high-carbon steel used in the making of cutting tools. In its annealed state it can be easily worked, but when hardened and suitably tempered it makes durable tools. It is often used for hand tools such as files and chisels, where their usage will not involve high temperatures that could draw the temper of the tool. Historically it was used for many tools such as drills and lathe bits, but today these are almost universally made of heat-resistant materials such as high-speed steel (HSS) and tungsten carbide. Examples of tool steels available to the model engineer are silver steel and gauge plate.

Many of the tools in the book have parts made from high-carbon tool steel, such as silver steel, which is supplied in a (relatively) soft condition and needs to be hardened before use. Carbon steel has many advantages for home-made tools: it is relatively easy to work, you can control the degree of hardness without expensive equipment and, once hardened, it can take a very sharp edge.

The basic process is as follows:

1. Hardening: this involves raising the metal to a high temperature, so that the steel transforms to its 'austentitic' form.
2. Rapid quenching: this 'freezes' the metal in the form known as martensite. At this stage, it is 'glass hard', but brittle.
3. Tempering: the metal is heated again, but to a lower temperature, reducing the proportion of martensite. This means it is somewhat softer, but much tougher.
4. Quenching again: the metal is cooled for a second time.

The details of the process can be varied to suit different steel compositions and to achieve trade-offs between hardness and toughness. Readers in North America will know hardening and tempering as tempering and drawing, respectively.

STEELS FOR HARDENING

The commonest grade of steel used in making small tools is silver steel, which is normally available as precision-ground round bar, and less commonly as square sections. It is normally quenched in water or brine (adding salt helps reduce bubble size during quenching). In North America, drill rod is a similar material and is available in a wider range of grades.

Gauge plate is an oil-hardening high-carbon steel that is available as precision-ground flats. It can be used for making form tools. It is also ideal for the making of small flat parts that need to be wear-resistant, such as ratchet clicks. The precise hardening requirements of gauge plate vary by manufacturer, but generally it requires a somewhat higher temperature than silver steel. It should also be quenched in clean oil, not water or brine, taking due precautions against any possible ignition of the hot oil.

These materials may come with detailed advice on heat treatment, but for non-critical uses you can follow the procedure of heating, quenching, pickling, tempering and cooling, outlined below. Obviously, you need to take sensible precautions to avoid any damage or injury from risks such as burns, fumes or fire.

First, make the part to size, leaving an allowance for grinding or honing after hardening, if that is required.

THE HARDENING PROCESS

Heating

Clean the parts. To reduce scaling during heat treatment and make the objects easier to clean afterwards, you can coat them in a slurry of chalk and water or methylated spirits, but for small pieces coating them in ordinary washing up liquid is effective. Place them on a piece of fireproof board or firebrick and use a blowtorch to bring them up to a bright red (the appearance of a boiled carrot is probably the most accurate reference). Work out of bright sunlight so you can easily judge the colour. It is difficult to illustrate these sorts of colours, but Fig. 4.1 is an attempt.

Another test is to see if it attracts a magnet. At a temperature of 700–800°C, carbon steel undergoes a transformation called decalesence, changing from various magnetic (ferrite, pearl-ite or martensite) body-centred cubic crystal structures into austentite, with a face-centred cubic structure that is non-magnetic.

For tiny parts, a miniature blowtorch is ample, and for most other parts, an ordinary gas-canister DIY blowtorch will be needed. Larger parts, however, will need a propane or butane gas torch running from a cylinder. Rapid heating helps minimize the erosion of sharp edges. A cyclone torch – a type of blowtorch that sucks in air near the base of the gooseneck and produces a whirling flame – produces a particularly fierce flame that wraps itself around the workpiece.

Fig. 4.4 Propane gas torch for intense heating of larger items.

You will also need some sort of hearth, best made from lightweight insulating blocks or firebricks. (Beware heavier types of firebricks that will suck all the heat out of your flame, rather than reflecting it back into the work.) It need not be complicated. For heating smaller work outdoors (which you may prefer to do), use some small pieces of vermiculite insulation board to surround the work and shelter it from any breeze.

Naturally, follow sensible safety precautions, such as allowing plenty of fresh air and making sure no flammable substances are anywhere close. Have suitable tools handy for picking up hot items. If using an oil quench for some steels, make sure it is in a metal container with an easily fitted lid to cap it if it starts smoking and to extinguish the flame if it catches light.

It may take a while to get the part up to temperature with a small torch. Once it is comfortably red all over, let it 'soak' for several minutes. A rule of thumb is

Fig. 4.1 Approximation of the colours of red-hot metals.

Fig. 4.2 A cook's blowtorch is ideal for heating very small items.

Fig. 4.3 A standard blowtorch.

Fig. 4.5 Using a gas torch with an improvised hearth.

five minutes for every quarter-inch of thickness for thorough hardening, but you need not worry about prolonged soak-heating a tool if it is not critical to harden it to the core.

With large parts, prolonged heating can cause decarburization – a form of burning of the steel, which will burn out the vital carbon from the cutting edges and diminish its hardenability, rendering it relatively soft. A layer of chalk slurry can help reduce this. Be aware that it is not necessarily a bad thing if the core of a large part does not get fully heated, as a softer core can make a tool tougher.

Quenching

Quench the parts by rapidly submerging them, or placing them carefully, into a volume of liquid sufficient to extract the heat without boiling away. Quench small tools end-first in clean water, with a gentle stirring action. Don't just shake it around in the water – if it is of thick section, it may well crack or distort.

Water gives a good quench for silver steel, but a 10 per cent brine solution is most effective and worth the effort for larger parts that hold a lot of heat. Gauge plate can be hardened with most oil – vegetable oil is pleasant – but this method requires a metal container with an easy-fitting lid, to prevent any risk of the smoke given off catching fire.

Rapidly quenching carbon steel when in the austenitic state converts it to martensite, which is an extremely hard form. The exact hardness depends on the amount of carbon in the steel and varies from 460 to 710 Brinell. Martensite is extremely brittle, glass

hard and unstable enough to shatter if dropped on a hard surface. It must be tempered before use by being re-heated and allowed to partially convert to softer, but tougher forms.

Pickling

After hardening and quenching, the resulting parts will look a bit miserable – rough and black. You can scrub them clean but soaking them for fifteen minutes in a pickle makes the task easier. Pickle is an acid bath for cleaning metal, as well as scale, and it can be used to remove spent flux and surface oxide after brazing or heat treatment. Many people use strong solutions of inorganic acids. A solution of citric acid (available in bulk from home brewers' suppliers) makes an excellent pickle and is far less risky to use. For small pieces of work, household limescale removers (usually citric or sulphamic acid) make a convenient pickle. In the past the metal was often immersed as soon as it had faded from red hot to black, causing the acid to boil and spit, but this is not advisable. Aside from the obvious hazards, this approach can cause unnecessary strains on the work.

Fig. 4.6 Citric acid makes a very effective but safe pickle for cleaning metals.

Tempering

After pickling, gentle rubbing with an abrasive pad should bring the metal to a matt, grey finish. Once the parts have been washed and dried, to prevent rapid rusting, they are ready for tempering. The controlled reduction of hardness makes the material less brittle, increasing its toughness and aiming for a compromise state that suits its application. (Incidentally, in the USA, the term 'tempering' is often informally used to refer to hardening, and 'drawing' is used to refer to the process known as tempering in the UK.)

When steels are heated in the presence of oxygen, a thin oxide film of a thickness related to the temperature forms on the surface. The colour of this film indicates the maximum temperature reached over a broad range, which usefully corresponds to that required for tempering hardened steel for various purposes. For example, springs can be tempered by heating until they show a dark blue colour, whilst many cutting tools only need to be heated until they take on a dark straw colour.

A clean surface is needed to get a clear view of the colour of the oxide film. There are many guides to the best tempering colours for different types of tool; Fig. 4.8 gives an indication of the colours and temperatures for different purposes, but clearly its accuracy is limited by the printing process. There is no substitute for gently heating a piece of steel and watching the colours develop for yourself.

Another alternative is to use a non-contact infra-red thermometer, if you have one. This is a relatively inexpensive

Oxidation Colour	Description and Temperature	Example Uses
	Light blue – 335 C (640 °F)	Parts where flexibility is needed, such as springs or ratchets.
	Dark blue – 310 C (590 °F)	Tools requiring toughness but not cutting ability, such as screwdrivers or tommy bars.
	Purple – 282 C (540 °F)	Punches, riveting tools.
	Brown – 260 C (500 °F)	Drill bits and striking tools, form tools and single point gear cutters.
	Dark-straw – 225 C (440 °F)	Most general-purpose cutting tools, scribers.
	Light-straw – 205 C (400 °F)	Reamers and precision cutters.
	Pale-yellow – 175 C (350 °F)	Scrapers and very hard tools for delicate jobs.

Fig. 4.8 Oxidation colours related to various tempering requirements.

Fig. 4.7 Approximation of the oxidation colours of steel.

device that can typically measure up to 380°C. Typically, the accuracy will be much greater than estimating from colour, but it will require a reasonably large target area to give an accurate reading. One way to achieve this is to place smaller parts on a piece of steel plate and heat them together.

You can re-heat the parts using the same method as you did for hardening, but apply less heat and go slowly. Bear in mind that the colours only appear when the steel is in ordinary air, so move the flame away periodically to check them. A popular approach is to heat the shank of a tool so that the colours 'run' towards the cutting edge. This leaves the shank softer, but tougher, which gives a more durable tool. For very small parts, make a small tray of sand in a metal lid and heat it from below.

Quenching is advised to stop small parts from overheating, but there is no harm in allowing them to air cool.

Another approach is to use an oven for hardening. This could be a temperature-controlled furnace, but a domestic oven is suitable for most parts. Gas mark 7 (220°C, 425°F) for twenty minutes works well for general purpose tools, giving a dark straw temper. You may need to experiment to find what temperature setting works best for your oven. This is a good approach for larger parts as it allows them to temper evenly right to the core.

Fig. 4.9 Non-contact digital thermometer.

ASSESSING HARDNESS

A hardness gauge can be used to assess the success of the hardening and tempering process. This is a circular disc with a series of pegs of known hardness arranged around its periphery. The pegs are used to scratch a sample, and its hardness is assessed as being between that of the hardest sample that cannot scratch it and the softest one that can. The hardness of the pegs is usually quoted in Brinell numbers.

Carbon-steel tools hardened and tempered in this way will always lose their hardness if re-heated to a higher temperature. For example, lathe tools that are run too fast will overheat, soften and wear rapidly. Be aware of this and do not expect carbon-steel tools to run as fast and as hot as those made from high-speed steel or tungsten carbide.

Dry and lightly oil tempered parts as soon as they are cool enough to handle.

HARDENING OTHER CARBON STEELS

High-Speed Steels (HSS)

In order to harden high-speed steels, they must be rapidly cooled from white heat. The use of oil or water baths for this purpose is impractical, and therefore a rapid blast of cool air is used. This requires the use of a special muffle furnace and the obtaining of annealed HSS. This may seem somewhat beyond the scope of the average hobby engineer, but it was not an uncommon practice in the past.

Low- and Medium-Carbon Steels

Low-carbon steels cannot be hardened in the same way, medium-carbon steels can be toughened for use in parts such as axles or heavy duties, but they will not become hard enough for cutting tools. You can buy such steels in a pre-treated condition, or harden them yourself by following the supplier's advice. Some grades of medium- and low-carbon steels are suitable for case hardening. There are two methods for case hardening. First, you can put the parts in a sealed metal box containing case-hardening powder, charcoal or other carbon-rich material, such as leather chips. The box is then heated to red heat for a prolonged period and carbon (along with nitrogen, if it is in the mixture) migrates into the outside skin of the metal.

Alternatively, small parts can be repeatedly heated to a bright red heat and plunged into case-hardening powder. Some case-hardening powders can release cyanide, so they must be treated with respect and the supplied safety instructions must be followed to the letter.

The 'case' of carburized steel produced by these processes can be hardened by plunging the red-hot part into water, after which it can be tempered. However, it is not unusual to leave the case fully hard as it is supported by the more flexible core of low-carbon steel.

ANNEALING AND NORMALIZING

The opposite of hardening is the process of putting metal into an annealed state, removing stresses and rendering the metal relatively easy to work. You may

need to do this if a hardened part needs to be re-worked, for example.

Changes in hardness of metals can be caused by either grain effects or changes in the crystal structure, or both. Most metals will harden when beaten, drawn or bent and high-carbon steels can be hardened by heat treatment. Annealing offers a way of making them easily workable. High-carbon steels are annealed by heating above the decalescence point (where they cease to be magnetic) and letting it soak through, followed by very slow cooling.

Normalizing, which simply corrects the grain structure to relieve stress in the metal, can be achieved more rapidly with cooling in still air.

OTHER METALS

Other types of metal cannot be hardened in the same way as carbon steels. One characteristic of many aluminium alloys is age hardening, caused by changes in the crystal structure of the metal. This may take place over days or hours, so any bending or forming should be carried out as soon as possible. Aluminium and its alloys can be annealed by heating almost to redness. This risks melting it, but the traditional way of avoiding melting is to mark it with soap, which chars at the annealing temperature. Note that some aluminium alloys have very complex hardening and annealing behaviour.

Copper and brass work harden easily if bent or hammered. They can be readily annealed by heating to redness, and rapid cooling can be achieved by quenching without causing hardening. If copper or brass are to be heavily worked, repeated annealing may be required.

5 Making Specialist Cutters

In a commercial production setting, the most valuable asset is time; the sheer number of items being made means that the cost per unit of specialist cutters becomes marginal. Even a jobbing workshop doing a specialist job is likely to be able to cost in any required tooling, especially if it may come in useful again in the future.

In hobby workshops, where the budget is tight and a cutter may be used only once, but time is cheap, there is great satisfaction to be had in the making of such tools. Indeed, as making your own cutters may take only an hour or two, it is often more efficient than waiting for a bought item to arrive. Cutters produced using hardenable carbon steels can be made very sharp and, although the need to avoid overheating and softening of cutting edges means they can only run at lower speeds, this is not a big issue for short runs of work.

This section will look at a selection of specialist cutters. Obviously, if you have an advanced tool and cutter grinder, you may have the skills and time necessary to produce top-quality tools from HSS blanks or silver steel. However, for the vast majority of cutters, the less sophisticated approaches described here will give perfectly acceptable results. You can hone edges with a slipstone or fine diamond 'files', but if you have a basic tool and cutter grinder of some kind, that will allow you to put a dead sharp edge on your tools.

There are a number of general principles that apply when making the cutters described below:

◆ Take care in hardening and tempering. It is better for a tool to be slightly over-tempered and run slowly than for it to be too hard and prone to breakage.
◆ Except for a few notable exceptions, provide relief behind cutting edges; sometimes you can leave a very thin witness when the size of a tool is critical.
◆ Ideally, sharpen tools by honing the front edge so that you do not affect sizing.
◆ Top rake is not essential, even for tools used in steel, although such tools will not cut as freely and should be used more slowly.
◆ Make sure shanks are as short and stout as possible; long, thin shanks are prone to vibration. For a tool such as a T-slot cutter, where a thin section is needed, keep this short and use full diameter for the main length of the shank.
◆ Except where you need a sharp corner, use rounded ones to minimize stresses wherever possible. This will reduce any chipping and wear of the cutting corners, as well as guarding against the likelihood of cracking in internal corners.

If you keep these principles in mind and work with patience, home-made tools can last for years and still give excellent results.

D-BITS

Of all precision tools, D-bits are among the ones easiest to manufacture from the scrap box. They are the true and proper destiny for the final 2 to 3 inches of any self-respecting bar of silver steel.

It is possible to take the manufacture of D-bits to a high level of craftsmanship. George Thomas, the renowned writer on workshop techniques, made a special jig to hold his D-bits while he milled them to the correct thickness. One design has a reduced shank, rather

Fig. 5.1 Using a D-bit to bore out a gear.

Fig. 5.2 A selection of shop-made D-bits.

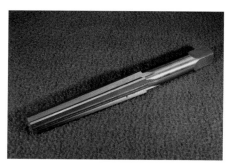
Fig. 5.3 Morse taper reamer.

Fig. 5.4 Shop-made taper reamers.

like a boring tool, to ease the flow of swarf. Nevertheless, for mere mortals, the simplest of designs will work (Fig. 5.2).

A D-bit cuts only on its end. The name comes from the cross-section of the tool, which must be marginally more (a few thousands of an inch, or less than a tenth of a millimetre) than a semi-circle to ensure that it does not cut on its edges. Because the D-bit is its own precision guide as it goes, it makes accurate-sized holes with a good surface finish. Some claim that a D-bit will produce better results than a reamer, which is not bad considering that they cost a few pence to make.

Manufacture is straightforward. Start by facing across the end of a piece of round silver steel. A very small chamfer all around at the cutting end will help self-centring, at the expense of a sharp inside corner to blind holes. Then grind, mill or file away slightly less than half of the diameter. Shape the end of the bar to provide clearance and form the cutting edge. An angle right across the end cuts a little more freely, but if the end is straight on the cutting side it will allow you to finish holes with a square bottom. Do relieve the non-cutting side of the end in any case.

After hardening and tempering to a dark straw colour, put a good finish on the end and the flat side of the bit with a slipstone – do not take any metal off the curved surface.

The tool shape is not good for heavy cuts, so a D-bit is used to finish a pre-existing hole. It does no harm to make a short pilot of the same nominal diameter as the bit, to get everything off to a good start. Beware overheating the tip, as this will draw its temper and it will rapidly become blunt. The secret is to keep the speed modest and use plenty of lubricant. Swarf soon builds up on the face of the tool, so it should be withdrawn regularly and treated to a dab of cutting oil.

A round-ended D-bit can be used to make round sockets and rivet sets. One which is flat across the end can be used to cut or renew valve seats.

TOOLMAKERS' REAMERS

The so-called toolmaker's reamer is an even simpler tool for making accurately sized small holes. In this case the tool blank is cut across at about 20 degrees, hardened, tempered and stoned to a good finish. Less robust than a D-bit, such reamers will take a very accurate final skim from an undersize hole.

Taper Reamers

Taper reamers are designed for finishing tapered holes to a standard size. Larger ones are available for machine tool tapers. Depending on the amount of material to be removed, the starting hole may be bored near to size (for just light finishing by the reamer); step drilled to an approximate taper; or for small sizes and shallow tapers, it may be a simple straight-sided hole. Taper reamers for making valve seats and similar purposes are easily made by turning a suitable cone in silver steel, then finishing it along the lines of a D-bit.

Where matching holes and tapers are needed, but the exact taper is not critical, it is possible to turn the required tapered parts, finishing by tapering a length of silver steel at the same setting. This can then be machine finished and hardened, in confidence that it will produce holes that are a close match for the other taper parts.

Taper Pin Reamers

A taper pin reamer is a slender reamer used specifically for making holes for taper pins. These metal pins (usually of hard brass or steel) have a gradual taper (often 1:48) and are used to secure a part to a round shaft. A diametric hole about the size of the small end of the pin is drilled though both parts and then shaped with a taper reamer. The pin is then hammered home and may be finished to length. Such fixings are semi-permanent and can be disassembled by driving the pin out from its small end with a pin punch.

Because of their small size and gradual taper, these reamers can be tricky to make. Rather than machining a flat along the tapered blank, it is easier to clamp it in a shallow groove on a piece of wood and carefully file it away to half its depth.

BROACHES

A broach is a long, edge-cutting, slightly tapered tool for enlarging holes. Unlike a reamer, which produces accurately sized holes, a broach is used on a cut-and-try basis to slowly enlarge a hole. Larger broaches can be made by milling a single flute along their length. Smaller ones can be made by filing three flats around a blank, but these should not be expected to cut very freely.

SLOT DRILLS

In most cases, it is probably best to use commercial milling cutters for general purposes as they are relatively inexpensive. HSS and carbide ones have a good life expectancy, can be run fast and are often practical to resharpen. From time to time, however, you may need to cut an accurately sized slot in one pass – and sometimes it may be of an unusual size. In these circumstances, making your own slot drills is a good option.

Select a piece of silver steel of the required diameter or turn down the end of a bar if necessary. Mill or file away two sides of the bar to create a tang of the required length and a suitable thickness – between a fifth and a quarter of the bar diameter is about right. File or machine relief along the two cutting edges, leaving a small witness. File a shallow V-notch in the end of the tool and add relief (Fig. 5.6).

After hardening and tempering, the tool can be honed along its cutting edges. Alternatively, if you have a suitable grinder, finish the cutting edges with it.

As always, a tool lacking top rake needs to be run slowly, but in use it should give good results.

BULL-NOSE AND ROUNDING CUTTERS

If you need to created rounded depressions or grooves, or round over the corners, you can make suitable specialist cutters by turning the end of a silver steel rod to the required shape, before proceeding as if you were making a slotting drill.

SPADE DRILLS

As with milling cutters, twist drills are difficult to make and cheap to buy, but sometimes you need an unusual size. Spade or spearpoint drills are an early design of drill, made by forging the end of a piece of carbon-steel rod to a diamond or spearpoint shape (when red hot). The end is then ground to diameter and the tool hardened and tempered. Cutting edges are ground on the end and the tool is finally stoned to a high finish. Lacking flutes, such drills do not clear themselves well, but they resist jamming and can be easily made in any required size. In small sizes, they

Fig. 5.5 Broaches.

Fig. 5.6 A crude slot drill.

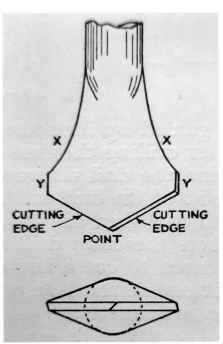

Fig. 5.7 Tip shape for spade drills.

are less fragile than tiny twist drills and are highly recommended for delicate jobs, such as drilling injectors.

COUNTERSINK BITS

Countersink bits are used to make conical depressions to accommodate the heads of countersunk screws or rivets. Usually, they have several cutting edges used for making such holes, which makes them very prone to chatter. Single flute countersinks work very well and can be made by turning a suitable angle on the end of a silver-steel rod. Normally, the countersink has an included angle of 90 degrees, but other angles are used. The bar can be milled or filed to half its diameter – rather more than a D-bit – for freer cutting. Such bits can also be used to chamfer holes.

Fig. 5.8 Commercial countersink bit.

SPOTTING DRILLS

These are stiff, short drills with narrower flutes than normal and a more acute tip. An alternative to centre drills, they are more convenient for marking the location of drill holes but cannot be used to prepare work for fitting to a centre as they have the wrong angle and no central pilot. They are not often used by model engineers, but they are considered superior to centre drills by many who have used them. A good substitute can be made in the same way as countersink bits, although they are likely to be somewhat longer and thinner.

CORE DRILLS

A core drill is a hollow cylindrical drill that cuts a circular slot and, if an object is drilled right through, it removes a plug of material (core). An emergency core dill can be made by drilling out the centre of a silver-steel rod, then milling or filing a series of teeth around the end. Without any set on the teeth, it will not be as efficient as a proper core drill and will need to be withdrawn regularly, perhaps every few seconds, to clear swarf. Such a tool may be useful, particularly when a shallow, circular slot is needed.

COUNTER BORES AND SPOT FACERS

A counter bore is an end-cutting tool with a pilot. It is used to make a suitable recess when the head of a screw or other fixing needs to sit flush with a surface. The pilot is engaged with an existing hole, so an accurately concentric bore to take the head of the fixing is easily made. A spot facer is used to form a flat pad around a screw hole in an otherwise unfinished area of a casting, or on a sloped surface. There is little practical difference between a spot facer and a counter bore, aside from the latter being produced in standard sizes to match available screws. There is no practical difference other than how far you drive them into the work, and the fact that the diameter of a spot facer is less critical.

Counter bores are a handy tool to make out of odd lengths of silver steel (Fig. 5.11). There are several methods. The easiest is to take a section of silver steel the same diameter (or larger) as the required tool. Turn the end down clearance size for the required fixings to make the short pilot, then if required narrow a second section to the counterbore diameter. The pilot should be about two diameters long

Fig. 5.9 Simple shop-made countersink.

Fig. 5.10 Centre drills (L) and spotting drills compared.

Fig. 5.11 Various counterbores and spot facers.

and undercut at the base – this is essential if the cutting edges are to be made long enough.

The cutting edges can be milled but can also be made by filing. You can file a series of notches to make multiple teeth, or file two sections of the larger diameter to make a flat cutter.

Now file the ends of the two cutting edges to provide relief (*see* Fig. 5.12). Don't worry about the exact angles but aim to leave a very small witness of the cutting edge. Be careful not to damage the pilot during this operation. There is no reason why you should not mill proper flutes on the tool, but the extra care and effort involved seems excessive for anything smaller than about half an inch diameter. One essential word of advice: make sure you cut the teeth the right way round; it is surprisingly easy to do them backwards! Compare with a drill to make sure you get the correct handedness.

If you file a series of notches around the periphery of the tool, you can create multiple cutting edges (*see* Fig. 5.13). It is an advantage if the teeth are not evenly spaced on this type of tool. It will be less liable to chatter as the irregular teeth will be unlikely to set up a vibration at a common resonant frequency. The tools should be hardened

Fig. 5.12 Counterbore with flat cutting sections.

Fig. 5.13 Spot facer with multiple teeth.

and tempered by heating to a 'boiled carrot' red colour, quenching, then tempering to dark straw.

If you find it difficult to file the teeth without damaging the pilot beyond usefulness, an alternative strategy is to drill a hole in the end of the bar. Fit a hardened pin in place after the tool has been sharpened – a force fit or ordinary cyanoacrylate glue is an adequate fixing, as stresses on the counter bore are minimal.

Another, more involved method, when you need a large counter bore, is to use a piece of steel the size of the required pilot. Chain drill and then file out a suitable slot near the end, then fit a cutter made from a rectangular piece of gauge plate, suitably notched on the cutting edge to locate on the pilot. A small wedge can be fitted in the slot behind the cutter to hold it in place. The cutter can be turned to exact diameter in place before it is hardened.

If the end of the shank of the spot facer is turned at an angle, it can be used to create conical depressions around a hole. Naturally, an ordinary centre drill can often be used for this purpose, but if you want a shallower recess – for example, to take countersunk screws – then this is a handy solution. A turned collar around the shank can be used as a depth stop

Fig. 5.14 Wood bits with adjustable countersink and depth stops.

or you could fit a moveable collar held in place with a grub screw, as in the countersunk wood bit in Fig. 5.14. The ability to drill multiple countersinks to identical depth gives a professional look to a row of fixings and done in this way is much easier than trying to use a standard countersink to the same depth for every hole. A strip of masking tape over the holes before countersinking will stop the depth collar marring the surface.

Another variation on this theme is the rose bit – essentially a counter bore with a hole instead of a pilot. A rose bit can be used to reduce the diameter of a short length of rod or even a screw, to create a spigot or pivot on the end. It has the advantage of being able to work on material too flimsy to turn down by normal methods. If the hole is the correct depth a rose bit can also be used to produce repeat parts with spigots all the same length, for example screws or rivets. Such a bit will not produce a perfect surface finish, but if you are going to finish by cutting a screw thread with a die, this is hardly critical.

STEP DRILLS

Step drilling is a technique of starting with a small hole and enlarging it in multiple stages. It is used where a lack of available machine power or the delicacy of a component mitigates against drilling a hole in one pass. It is also a good approach for making larger holes in sheet metal, with less tendency for the hole to assume a lobed shape. A step drill in the form of a stepped cone with straight or spiral flutes is used for drilling large, accurate holes in sheet metal in a single pass. Due to the large increase in diameter between steps, the work needs to be held firmly and ideally supported by backing material to prevent distortion.

Commercial step drills are often rather costly (being hefty lumps of HSS) and have a distinct disadvantage in that by design they only cover a discrete series of diameters. If your step drill ascends in even numbers of millimetres, how do you drill a 21mm hole or one that is measured in imperial?

It is straightforward to turn your own step drill from silver-steel bar. A standard 1in bar can cover a useful range of metric or imperial sizes. Note that the steps are not flat but should be slightly angled at 5 to 10 degrees to ease the start of each new step. A single flute to create the cutting edge is best produced by milling and need not be much

deeper than the steps. The very end of the drill can be milled straight across on each side and then sharpened like a spade drill, rather than fluted. Alternatively, you could use a pilot drill or even glue a suitable stub drill in the end of the tool after hardening.

In theory, such a large piece of silver steel should be heated for a long time before quenching; in practice, it is only necessary for the outside of the tool to harden. If you place it cone upwards and apply heat direct to the steps, you can keep the period of heating shorter and reduce the risk of decarburizing the cutting edges.

Rather than attempting to make a spiral flute, using a straight flute makes it easier to hone them by polishing the flat surface with a slipstone.

CONE DRILLS

When the exact size of a hole is less important, and a slight taper does not matter, a cone or taper drill – a conical single or double-flute drill (Fig. 5.16) – can be used for opening up holes in sheet material to any size within its range. Because these drills have a very steep taper, they do not make parallel-sided holes. They are best suited to

making holes for electrical controls and sockets in panels and other non-critical applications. They are a simpler alternative to step drills and making them is literally the same, but without the steps.

T-SLOT CUTTERS

Machine tables, jigs and other devices often use T-slots, shaped to allow the use of T-nuts for clamping items in place. Caution should be taken not to overtighten fixings, as T-slots can be damaged in this way. T-slots should be kept clean and free of swarf as far as is practical. It is an advantage, but not essential, for the various machines in a workshop to share a common T-slot size. Certainly, if you are making your own T-slots, it makes sense to match sizes for which you already have a stock of T-nuts and clamps.

T-slot cutters are not used to make the entire slot, only the broad section at the bottom of the slot, and an initial plain slot should be end-milled to full depth, or even slightly deeper, first.

To make a T-slot cutter, turn the end of a bar of silver steel to match the full width of the slot, then turn away a section to slightly less than the width of the upper slot to leave a disk at the end. Teeth can then be milled or filed around the periphery of the disk before hardening and tempering. If the teeth have relief behind the cutting edges and are angled slightly, the cutter will cut much more freely. If the face of the cutting edges is placed slightly behind a radius line from the centre of the cutter, it will have a little top rake and cut better.

These cutters should be treated gently, as they perform quite a demanding task;

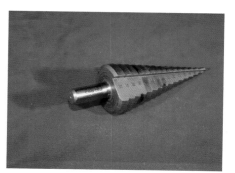

Fig. 5.15 Step drill.

Fig. 5.16 Cone drill.

Fig. 5.17 T-slot cutters.

Fig. 5.18 Dovetail cutter.

do not run them too fast, use cutting oil and clear swarf regularly.

WOODRUFFE CUTTERS

Woodruffe keys are semi-circular keys that fit in a matching pocket in a shaft. Such a key normally fits entirely within the boss of the surrounding component that has a normal straight keyway. A Woodruffe cutter is a circular cutter that resembles a thin T-slot cutter and can be made in the same way.

Woodruffe cutters are also useful for cutting grooves and are used by some locomotive builders for grooving connecting rods. Ideally, such grooves should have rounded, not square, corners; they are an ideal application for shop-made tooling.

DOVETAIL CUTTERS

Dovetail cutters are typically multi-flute rotary cutters in the form of a truncated cone, used to cut dovetails for machine slides and similar purposes. The commonest dovetail angle is 60 degrees, but other angles such as 45 and 75 degrees may be encountered.

The basic form of a dovetail cutter is relatively easy to machine but beware

Fig. 5.19 Shop-made and commercial cutters compared.

of making the part joining the cone to the shank too narrow. A freer-cutting tool will result if the end of the cutter blank is slightly recessed before the teeth are cut. As with a T-slot cutter, the teeth can be filed or milled.

The long edges of dovetail cutters mean that it is particularly important not to overload them and to ensure a good finish. Fig. 5.19 compares commercial T-slot and dovetail cutters with rather smaller shop-made ones.

SPECIAL CUTTERS

Making your own tooling gives you an exceptional amount of flexibility. If you need an awkward shape or a groove of an unusual size, for example, it is a fairly quick process to make a simple cutter to do the job. The cutter in Fig. 5.20 was

made just to cut some slots in the sides of a tool-room vice to provide more flexibility for clamping it down on the mill table. It was made from a slice of silver steel with a central hole, then a ring of holes was drilled around the end, and these were sawn and filed to create the cutting teeth. It did the job of cutting two 100mm long slots easily and has since been used for several other jobs – keep your 'special' cutters safe, as you never know when you might use them again!

The special cutter in Fig. 5.21 was made to a design by LBSC and was used to cut three accurately spaced slots as valve ports in steam engines. The starting point was a blank turned as three suitably dimensioned disks, which were then milled away to form the three cutters.

Fig. 5.20 Slotting tool fitted to a simple shank.

Fig. 5.21 Custom tool for cutting valve ports.

FORM TOOLS

Many of the tools described above can be used as form tools; that is to say, they can be used to create a profile that matches their shape. Examples include bull-nose cutters, gear cutters and dovetail cutters, and even slot drills count.

For more complex shapes a practical way to make form tools is to machine or file the negative of the required shape in a piece of gauge plate, not forgetting to ensure there is some relief beneath the cutting edge. Remember that gauge plate is an oil-hardening steel, and this requires a slightly different process. Once hardened and tempered, such tools can be used to shape a workpiece held in a lathe or fitted to a shank and used as a fly cutter.

The often-extended cutting edges of such tools mean long lengths of cut, so they are vulnerable to chatter. Because of this, they should be used in as rigid a set-up as possible, at low speeds and with adequate lubrication.

The example in Fig. 5.23 was made by drilling a 10mm hole at an angle of about 5 degrees to create the relief, before the unwanted parts of the strip of gauge plate were cut away and the tool hardened and tempered. This tool

Fig. 5.22 Turning the blank for the valve port tool.

Fig. 5.23 Form tool for turning brass balls.

Fig. 5.24 Governor using form tool-turned balls.

was then used to produce governor weights for a model from brass bar.

PUNCHES

Leather Punches

Leather or hollow punches are used for the obvious purpose of making holes in leather, but they can also be used on card, plastic, thin timber and very thin, soft sheet metal such as aluminium from drinks cans. With care, it is possible to produce thin sealing washers by using two punches in succession.

The standard design (*see* Fig. 5.25) has a 'stirrup' shape so that the cut circles can be removed easily. You can make your own by drilling a silver-steel rod at one end, then a cross-hole right through, and turning a bevel on the end. Round

Fig. 5.25 Leather punch.

off the top so you can strike it without raising a burr. It is a good idea to poke the cut material out with a wire every few holes, otherwise it will block solid.

Sheet-Metal Hole Punch

It is often a requirement to make one or more accurately positioned round holes in sheet metal. Anyone who has ever tried to drill large holes in sheet brass will know that they are more likely to come out triangular than round! One alternative method is to drill smaller holes and use taper hand reamers to enlarge them to size. However, although this makes round holes, it is hard to get them accurately positioned. Instead, a circular hole punch will do the job well. It also provides a good introduction to hardening silver steel, a skill that will be

Fig. 5.26 Sheet-metal hole punch.

required for some of the other projects in this book.

The tool was originally produced to make a line of nine neat 13mm diameter holes in a sheet of 20-gauge brass for an upright boiler's firebox. With the size easily modified, these punches are ideal for making mounting holes for many electronic components such as sockets, potentiometers and switches.

These punches are fairly costly to buy but are not difficult to make. The raw materials for the example in Fig. 5.27 were short pieces of 25mm and 13mm diameter silver steel, an offcut of 19mm steel hexagon and a 25mm M6 cap screw. The exact dimensions are not important, except that the punch needs to be dead on size and the hole in the die slightly oversize.

The body of the punch is a 13mm length of 1in diameter silver steel. This needs to be drilled and (ideally) reamed 1/2in – an easy sliding fit on the smaller stock. Face both sides of the resulting ring to give sharp edges to the hole but break the outer edges of the ring to make sure the punch is comfortable to handle.

The punch itself is made from about 7mm of 13mm diameter silver steel. Measure the screw and drill this a close clearance fit on the cap screw. Do not drill it 6mm for an M6 screw – metric screws are significantly smaller than their nominal diameter, so 6mm will be too loose and the punch will be liable to jam. Hold the punch between soft jaws in a vice and, using the back of a half-round file, create a groove and widen it to the full width of the punch body. This will create a curved edge so that the punch has a 'scissors' action, rather than trying to shear out a complete circle in one go. An alternative is to file the end of the punch like a pitched roof.

However you shape it, make sure that the cut starts at two opposite points, as this will help make sure the punch enters the body truly. Carefully clean off any burrs raised on the punch with a slipstone, but take care not to blunt the cutting edge.

The final part is the nut. Make a spigot on the end of the steel hexagon that is a good fit in the body of the punch, no more than 5mm long. Slightly relieve the edges of the hexagon around the spigot. At the same setting, drill and tap for the cap screw. If you do not have any suitable hexagon material, you can use round bar, but you will need to file or mill spanner flats on it. Use your tailstock chuck to guide the tap to ensure concentricity. It is vital that both the hole in the punch and the thread in the nut are exactly in line when everything is assembled, otherwise the punch will catch on the edge of the body and jam or chip. Finally, part off to length.

You can now do a trial assembly, and even try out the punch on a sheet of card, but it should not be used on metal until it has been hardened and tempered (*see* Chapter 4).

To use the punch, the first step was to mark out and centre pop the locations of the hole centres. Using a wooden

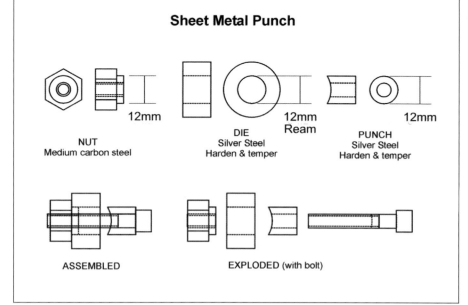

Fig. 5.27 Construction of the hole punch.

Fig. 5.28 Parts of the hole punch.

Fig. 5.29 Assembly of the punch.

Fig. 5.30 Waste metal from hole cutting.

backing board, drill each hole 6mm, using a clamp to ensure the sheet cannot spin as the drill breaks through. Fit the punch to each hole in turn, pulling it through from the 'good' side of the sheet. A small amount of oil helps make sure everything goes smoothly. It is easier to hold the hexagon nut in a vice than to wrestle with both a spanner and an Allen key. The holes should come out with clean edges that need no further finishing. The quoted sizes allow the punch to be drawn right through the brass sheet, which means it comes free and does not need to be pulled off the punch, minimizing distortion. The rings (Fig 5.30) may tend to stick in the body of the punch, but it is easy to tap them out. You could step the body with a larger diameter underneath so that the rings drop out, but if you do so, remember to make the spigot on the nut larger to match.

Naturally, the sizes of the punch can be varied to suit the job in hand. If you want to make really large punches it may be worth considering case hardening them (*see* Chapter 4) rather than using large-diameter silver steel.

6 Workholding Aids

Anyone who aspires to accuracy in their work will find that they need a good selection of tools to secure pieces of work while operations are performed on them. This includes not only the obvious vices, but also clamps, jigs and fixtures. You will find that even hand work, such as filing and sawing, will be improved by an order of magnitude if the piece is held securely and presented so that you can work comfortably. Some of the most important aids to workholding include blocks, clamps, vices and parallel bars.

BLOCKS

V-Blocks

V-blocks are precision-ground metal blocks with a large V-shaped notch in one side, often hardened and used for supporting work for machining and marking out. They are usually supplied in matched (typically numbered) pairs so that workpieces can be supported with a block at each end, allowing through drilling between the blocks (*see* Fig. 6.1). V-blocks are ideally suited to holding round work but may also be used to hold other shapes. The slot at the bottom of the V ensures that square-cornered work will sit fully in the block. They are often, but not always, supplied with matching clamps

Fig. 6.1 A matched pair of V-blocks.

Fig. 6.2 V-blocks with fitted clamps.

to facilitate workholding. Fig. 6.2 shows how such clamps are used; note also that these blocks have two V grooves of different sizes. Both styles of block can be clamped down with milling clamps or held in a vice.

1-2-3 Blocks

A 1-2-3 block is a simple fixture that assists in setting up and laying out,

consisting of hardened and ground rectangular blocks with faces in the ratio 1:2:3. Inch dimensions make a convenient size. The blocks are drilled to allow flexibility in clamping and use as packing.

Stevenson's Blocks

A variation on the 1-2-3 block theme is Stevenson's Metric Blocks, which are available in various sizes. They have faces in the ratios 1:2:4. These blocks have sixteen holes, half of which are tapped and the rest clearance size. This allows them to be clamped together to create bespoke jigs and fixtures or even temporary angle plates.

Bench Block

A bench block at its most basic is simply a chunk of steel plate used as a basic miniature anvil. A proper engineer's

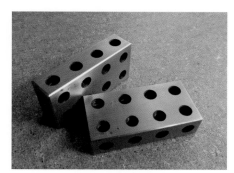
Fig. 6.3 A pair of Stevenson's blocks.

bench block is more sophisticated and includes holes to suit various sizes of material and a V-groove for positioning work for cross-drilling, punching out rivets or all sorts of other duties, such as bending wires and helping hold small-diameter rods when cutting them with a junior hacksaw. Producing one is a nice little project that will make good use of a short offcut of relatively large-diameter steel bar.

The bench block in Fig. 6.4 was made from a slice of 75mm (3in) mild steel. The milled V-groove has three holes of different sizes to facilitate cross-drilling of round bars. The groove is made by supporting the block at a 45-degree angle and using an end mill. Do not forget to use a small end mill to cut a thin groove at the base of the V.

The other holes are a wide range of sizes. Underneath, the central portion is recessed to make sure it sits flat on a level surface (*see* Fig. 6.5), and a knurled pattern around the edges makes it easy and pleasant to hold in position. When you make your own bench block, feel free to change the dimensions and features to suit the sorts of jobs that you plan to undertake. Take the time to finish it nicely – at the very least it will make a handy workshop paperweight

Fig. 6.5 The underside of the bench block is recessed.

for all the scraps with your plans and lists scribbled on them!

PARALLELS

Parallels are precision-ground metal bars that can be used to space and position work accurately. Usually, the work is clamped down or against the parallels, but the holes in the parallels shown in Fig. 6.6 allow bars to be threaded through them, which can then be borne down on by a clamp. Inexpensive alternatives to parallels include old ball races and lengths of accurately ground square stock (Fig 6.7).

Wavy parallels are paired sinuous bars from spring steel that have been ground parallel on their two long edges. As well as serving as ordinary parallels, they are

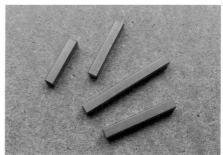

Fig. 6.7 Pieces of ground square stock used as general-purpose parallels.

easier to stand upright so they can be used as precision spacers. They can also be used as light-duty spring clamps.

VICES

Soft Vice Jaws for Round Work

One of the great advantages of 3D printing is that you can produce jigs and fixtures without a significant investment in material. If they do the job, then it is not a big issue if they end up straight in the recycling bin afterwards. That said, there is no reason why some devices may not have an extended life, depending on the job they are asked to do and the care with which they are used.

The set of soft vice jaws in Fig. 6.8 is a good example. It pops straight between

Fig. 6.4 Bench block.

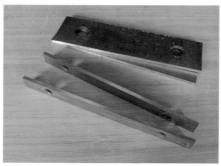

Fig. 6.6 Accurately ground parallels.

Fig. 6.8 Design for 3D-printed vice jaws for round items.

Fig. 6.9 3D-printed jaws in use.

the jaws of a 100mm (4in) width vice and allows small round parts to be held for machining or handwork (Fig. 6.9). Obviously, the security of grip is limited, so perhaps this solution is not suitable for heavy milling tasks. It is, however, ideal for putting a thread on brass rod or holding a part for filing or sawing, for example. It also has the advantage of being kind to more delicate items such as fine brass screws, which could be damaged on their threads by metal jaws. With smaller holes it would suit a clockmaker who needs to hold small screws or pivots for polishing without marking finished surfaces.

How long the jaws last will depend on how tightly the items need to be gripped, and how much care is taken not to machine or cut into them.

It is not straightforward to decide the printing parameters for a job like this. A high level of fill, say 50 per cent, gives strength, but you do not want 100 per cent fill, as some resilience is useful. A wall thickness of 2mm is higher than normal but helps make sure the grips do not distort too much. These settings result in a relatively long print time, so you could go for relatively thick layers of 0.2mm to speed up the print but also to give a more ridged and, hopefully,

a grippier surface. The item is best printed upside down, to avoid the need for support of the locating flanges. This also ensures a nice smooth top surface.

Naturally, this design can easily be modified to suit different vices. It is also possible to use it to hold all sorts of weird and wonderful shapes, simply by creating 'blank' jaws and cutting out the profile of whatever it is you want to hold.

Vice Spiders

A recurring problem for lathe work is setting work close to the front of the jaws without getting it slightly out of line. The obvious solution is to pack it out with parallel, but if you have it set well enough, the parallel is hard to remove without scratching it, while leaving it in place is potentially hazardous. A 'chuck spider' – a three-armed device with parallel front and rear faces – is a great aid to setting up work in this way.

The spider can be sawn or milled from a flat sheet, but those who have one may take advantage of the ability of a well-levelled 3D printer to produce accurately parallel surfaces of almost any shape to produce custom spacers (Fig. 6.10). These can be shaped so that

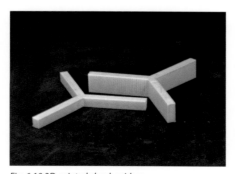

Fig. 6.10 3D-printed chuck spiders.

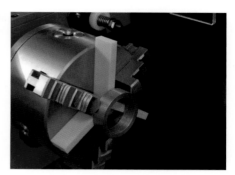

Fig. 6.11 Chuck spider in use.

they cannot escape the jaws; even if they were to break free, the light plastic would be a lot less dangerous than a flying metal bar! Another advantage is that the spacers can be 'sacrificial' if you need to drill or bore right through the work. They are very simple to use, and very effective.

The normal approach is this copy of a popular design often made in metal – a three- or four-legged 'spider' to suit the chosen chuck – but any weird or wonderful cross-section can be used, as long as you make sure it is too wide to slip sideways between the chuck jaws and big enough to comfortably bridge the chuck bore.

This is an example of an accessory that should have a long life. Unless you drill or bore into it, it will last indefinitely, although it should preferably be stored away from sources of heat.

BUSHES AND MANDRELS

Staying with the theme of holding work accurately, one myth of the modern workshop is that the three-jaw chuck is inherently inaccurate. The story goes that only the best can be relied upon and, if you overtighten such a device just once, it can lose its accuracy for ever. However, tests with a dial gauge along the length of a 6-inch piece of silver steel chucked in the three-jaw chuck that came with an inexpensive mini-lathe showed that it was still no more than 0.025mm (0.001in) off centre, 100mm (4in) from the chuck. Others have reported similar results. Admittedly, the chuck does need to be freshly cleaned to turn in this sort of performance, but experience in resetting work does suggest that even an inexpensive chuck can have a good level of basic accuracy. This can be the case after several years of use, so it is worth checking your own; it may exceed your expectations. Incidentally, the late George Thomas performed similar tests on a number of drill chucks, and was surprised at their accuracy.

Even so, at times you may want to be absolutely sure that work is held truly concentrically. A simple way of achieving this is a split bush, bored or drilled and reamed *in situ* to a good fit on the work, but drilled and reamed if the bore is small. A 'top hat' shape is convenient, as it can be positively located against the chuck jaws. If the slit is

made opposite to jaw number 1, you will have a sporting chance of locating it accurately again in the future.

Almost any bit of scrap can be used to make such a bush, but brass is a good material, as it is easy to work accurately and to a good finish, while having a degree of 'springiness'.

For hollow work, a mandrel that fits inside the work is needed instead. Turned in place, such a mandrel will be 100 per cent accurate. If you turn them from offcuts of hexagon material (marked for number 1 jaw), they can be re-used for less critical applications.

This can be turned to a tight fit, relieved to give a slight taper at the outside end. This is the most accurate solution, but the danger is always there that the work will skid and be scored on its bore. A threaded end for a nut and washer, or a screw and washer, can be used to hold most work secure. Another alternative when you need to be able to turn the whole face of the work is to thread the mandrel undersize for a large screw (just use the tip of a taper tap), then split it with a neat saw cut; once fitted inside the work, it can be

Fig. 6.14 A selection of different mandrels made for various workpieces.

expanded by fitting a suitable screw. Fig. 6.14 shows a selection of mandrels made for different tasks.

CLAMPS

L-Clamps

These simple little clamps are ideal for holding small or thin pieces of work very securely. They are also suitable for holding in position those vices that have a narrow groove parallel to their bottom surface.

They can be made very quickly if material of a suitable L-section is available. Those in Fig. 6.15 were made from some small pieces of scrap malleable

Fig. 6.12 Split collet for holding round stock accurately centred in a chuck.

Fig. 6.13 A pair of stub mandrels.

Fig. 6.15 L-clamps from malleable cast iron.

cast iron that only needed slicing into suitable pieces. A clearance hole for a fixing bolt was then drilled. The short tip of the 'L' should be filed into a slightly convex shape so it can rest evenly and does not have a sharp edge pressing into the surface below. Another suitable source of material is stout angle iron, with one leg cut down.

The proportions of the clamps can be varied to suit the material available. For maximum clamping pressure, have the hole near the clamp point, and round off the end of the short leg of the 'L' to avoid marking the surface the clamp is tightened down on. The one thing to watch out for is making the clamps too long – it is amazing how rapidly clamps will start to flex as you make them longer. The amount of flex increases with the cube of the length, so if the length of

the clamp is doubled, it will take only one-eighth of the clamping force!

It is worth considering making up fresh clamps like this for any odd jobs that come along, so that you build up a selection of different devices to tackle all sorts of challenges. It is also a good idea to make some T-nuts to suit your various machines with suitable fixing bolts and keep these with the clamps. High-tensile steel bolts should be used, but using low-carbon steel for the T-nuts will reduce the wear and tear on T-slots.

If you need really strong clamps and know the height at which they need to be used, an alternative approach is to make a block as in Fig. 6.10. The lip at one end of the base acts as a pivot, and the short protrusion does the clamping.

The T-nuts being used with the clamps are also home-made. They are

simply milled from steel bar, drilled and tapped to suit. M8 and even M6 bolts may be perfectly adequate on your milling machine, and much easier to use than the huge M10 fixings that come as standard.

G-Clamps

G-clamps are one of the commonest types of clamp. Their main advantage is the ability to apply a relatively powerful grip over the whole range of capacity for their frame. Unfortunately, they have a few drawbacks for engineering purposes. The main problem is that the screw adjustment gives them a tendency to twist workpieces out of alignment or causes them to 'walk' across the job as they are tightened up. This is rarely an issue on a rough timber surface but can be very frustrating when trying to grip smooth steel. A second issue is a tendency to loosen off in response to vibration, which makes them unsuitable for holding material for machining.

On the other hand, well-made small G-clamps are handy for holding small objects for jobs such as gluing or

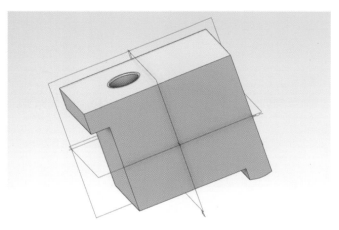

Fig. 6.16 Shop-made T-nuts.

Fig. 6.17 Design for vice hold-down clamps.

Fig. 6.18 A G-clamp and speed clamp compared.

Fig. 6.19 A precision G-clamp.

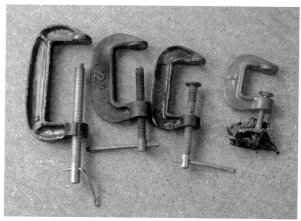

Fig. 6.20 A selection of abused G-clamps.

One often-overlooked aspect of these clamps is that, if the two heads are reversed, they can be used as spreaders to push things apart.

Toolmakers' Clamps

The traditional engineer's or toolmaker's clamp solves the issues with speed clamps and G-clamps, having no tendency to twist and being able to apply very great pressure. The downside is that they can be rather fiddly to use, as they have two separate adjusting screws. One of these is situated about half-way along the clamp's bars, the other is at the far end from the clamping section. The way to set a toolmakers' clamp is as follows:

- Wind the rear screw out as far as is required to allow the clamp to close sufficiently.
- Adjust the central screw to set the two clamp bars the right distance apart; it can help to hold the clamp across the work and 'feel' when the bars settle into position. Set the bars

soldering. G-clamps can also prove useful for holding items that are going to be exposed to heat, as when welding or brazing, or for holding the two halves of a mould together. Clamps used in this way can become tatty very quickly, but it is worth keeping a stock of cheap or damaged ones for this sort of task.

Speed Clamps

Speed clamps have two heads – one fixed, one moveable – on a steel bar. They overcome the main issue with G-clamps by not having any tendency to twist and are particularly useful for clamping parts for gluing in an exact location. Typically, they have a 'trigger'-operated ratchet action, but their gripping power is usually rather modest. As they often have plastic parts, they are unsuitable for use with heat.

Fig. 6.21 Speed clamp arranged for use as a spreader.

so they hold firmly in position but do not overtighten them.

◆ Move the rear screw in again so that it wedges the rear of the clamp apart slightly, increasing the pressure on the work.

◆ Depending on your requirement, you can either try to set the clamp bars parallel or allow them to tip up slightly to maximize pressure at the jaw tips. It only takes a little practice to make setting up the clamps reasonably easy.

A toolmaker's clamp is easily made. There are no magic sizes, but it is important to achieve a sensible balance between the size of the screw threads and the bars. To ensure they assemble easily and work well, it is best to start by clamping the two bars together, drilling the middle hole through at tapping size, and the back hole through one bar and half-way through the other. The holes in the bar with two through holes should be threaded, while the middle hole in the other bar is opened up to clearance on the screws. The 'half hole' at the back for the adjusting screw can be opened up to clearance size, but a neater solution is to turn a short section of thread off the end of the adjusting screw.

Fig. 6.23 Miniature toolmaker's clamp made for a delicate job.

It is often easiest to use studding or 'allthread' to make the screws and fit knurled or cross-drilled knobs to facilitate tightening. For the best results, use a high-carbon steel and harden and temper to a deep blue. Leaving the clamps very hard not only makes them

Fig. 6.24 The screws on this clamp are 'allthread' with brass knobs fitted.

brittle, but also makes them more likely to slip. Alternatively, case harden the clamps (*see* Chapter 4).

Fig. 6.25 is an example of how a toolmaker's clamp can be used creatively, exploiting its capacity to have its jaws set parallel.

FLANGE DRILLING JIGS

The relatively simple jigs in Fig. 6.26 are a good example of how a simple device can make an awkward exercise much easier. Small steam fittings for model engines are awkward shapes to hold, making it difficult to drill their flanges accurately (Fig. 6.27). The easy solution is to make recessed and drilled disc templates from silver steel, hardened and tempered, as suggested by Tubal Cain. These have the advantage that they allow the easy, accurate drilling of the flanges even after they have been attached to the pipes.

Fig. 6.26 Flange drilling jigs.

Fig. 6.27 Steam fitting in the drilling jig.

Fig. 6.22 A pair of toolmakers' clamps.

Fig. 6.25 The clamp used with a three-jaw chuck and rotary table to hold a part for rounding its end.

7   3D-Printed Tools

One of the latest technological devices to find its way into hobby workshops is the 3D printer. Initially dismissed as a novelty by many, its usefulness really depends on how inventive you are. Some of the uses found for 3D printing in the workshop include the following:

- project boxes;
- connectors and fixings;
- small tools;
- jigs to hold awkward parts;
- custom gear wheels;
- patterns for making castings.

3D printing is now becoming a mature technology, and the latest printers are very easy to use, although self-assembly kits are still popular as a relatively inexpensive way to get into this fascinating aspect of hobby engineering.

TECHNOLOGY AND MATERIALS

The first thing to understand is the technology used in these devices. Most 3D printers use a process called 'Fused Deposition Modelling' (FDM), also known as 'Plastic Jet Printing' (PJP) or 'Fused Filament Fabrication' (FFF). This technique involves layers of a specific material being laid down repeatedly until the object has been created. The cross-section layers are formed by a continuous length of plastic being ejected

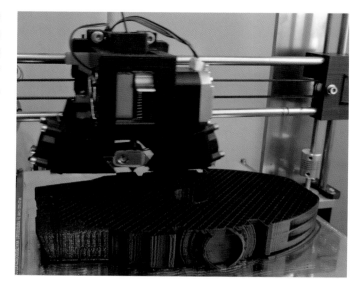

Fig. 7.1 A fused filament printer lays down plastic in thin layers.

with a heated nozzle. The plastic then quickly hardens. Other printers use 'Stereolithography' (SLA), in which a laser solidifies layers of photosensitive resin. At first, only FDM machines found a wide applicability in home workshops, but resin printers are becoming more and more accessible.

Of great importance is the printing material itself. Most printers use a spool of filament in either PLA (polylactic acid) or ABS (acrylonitrile butadiene styrene). These are thermoplastics, which means they can be heated to become flexible and revert to being solid once cooled. ABS material is flexible, strong and has a high temperature resistance but tends to produce unpleasant fumes during printing. PLA can produce thinner

layers, sharper corners and is available in a wider range of colours, making it the preferred choice of schools and hobbyists. The advent of so-called 'PLA-plus' has seen a big increase in the ease of printing reliably with PLA. However, there are many other materials available, including nylon and PETG.

CHOOSING A PRINTER

When choosing the right printer for you, you need consider various attributes. What is its print area? The bigger it is, the bigger the objects that may be created. Look at the printing speed as well, but be aware that the type of material used and the complexity level of the print can affect speed more than a particular printer's capacity. Also, think about the layer resolution. Some 3D printers allow you to enter the value of the layer resolution, whereas others provide options. This will affect the outcome of the final printed object. Very fine resolutions are agonisingly slow but can give very good results. Lastly, the extruders of the printer are a key attribute of the device. These allow for the melted printing material to be ejected, so more than one extruder will allow you to print in different colours and different materials. Ideally, extruders that allow easy filament changing can make life a lot easier.

3D printers also offer many different non-essential features, which can make the printing experience easier and more enjoyable. Self-levelling eliminates some of the most awkward parts of setting up the printer. Most will come with options to allow for wireless connection or an SD slot and the option to connect directly to a laptop. More informative LCD displays that relay information back on the printing progress can help with managing your printing and with fault-finding.

Most 3D printers come with their own software that can open STL files, which is the standard file format used to print a 3D model. The software allows you to choose print settings, layer thickness and fill types. However, to design and create your 3D models, you will need 3D CAD software.

Settling on a preferred 3D printer will largely depend on your budget. Today, most people choose an off-the-shelf printer, but building from a kit can be rewarding, save money and give results that are just as good – it can also make customization and improvement

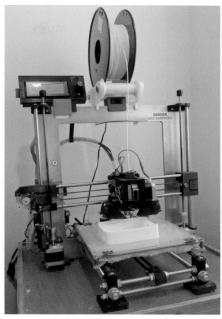

Fig. 7.2 A Prusa-type 3D printer.

easier. Ready-made printers start from amounts in the hundreds of pounds, going into thousands of pounds for the higher-end type of products. It goes without saying that, the more expensive the printer, the more features it will have. Essentially, its price will reflect its overall output quality and functionalities.

COMPUTER-AIDED DESIGN (CAD)

The ready availability of computers and computer-aided design (CAD) packages means that most hobbyists now reach for the mouse rather than a pencil. If you are going to design your own 3D-printed objects, using some form of CAD is essential.

Gone are the days when hobby CAD meant using a program really meant for drawing; there are many affordable and even freeware programs readily

Fig. 7.3 Parts designed in TurboCAD.

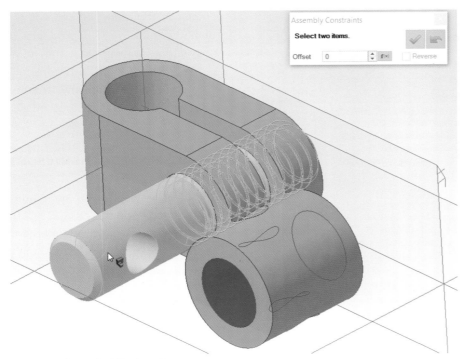

Fig. 7.4 Parts designed in Alibre Atom 3D.

Fig. 7.5 3D-printed knob to fit commercial bolt.

Fig. 7.6 Underside of the bolt showing bevel.

available now. While this brings many advantages, some of the most useful are automatic dimensioning, which reduces errors, and the ability to easily move around parts and combine them in two and three dimensions. Some packages even let you 'model' the movement of mechanisms.

Freeware CAD programs include FreeCAD and LibreCAD. Fusion 360, which was a popular free program, is now free only for an extended trial period, after which you must buy a licence. TurboCAD, often at the level of 'TurboCAD Deluxe', is popular with many hobbyists and works much like the industry-standard AutoCAD. Another favourite paid-for program is Alibre Atom 3D.

Many people who learned traditional technical drawing start with 2D drawings that they then convert into 3D models, but complete beginners are advised to start straight away by making 3D objects. This avoids having to learn everything twice, and it is much easier to make 2D drawings from a 3D model than the other way around. Naturally, the detail of how you produce a model is different for each package, but a general principle is to create 3D blocks, perhaps by extruding 2D shapes, and then sculpting them by either merging additional features or 'subtracting' other objects. For example, if you need to create a threaded hole through a block for a screw, you can line the two objects up and subtract the screw from the block.

3D-PRINTED KNOBS

To find an example of a useful project, just look around your workshop. How many of your machines have parts held in place by cap-head screws that send you scurrying to find an Allen key for the most trivial of purposes? Examples might include gear and belt covers, and lathe splashbacks; one particularly annoying one is the large blade cover on a bandsaw. None of these requires a lot of torque to keep closed and a simple thumbscrew removes the need for an Allen key.

Whilst 3D-printed female threads can be reliable, long male threads are less so. Rather than print the complete part, one solution is to print plastic knobs into which suitable screws can be fitted. The basic knob is made around a simple cylinder, pierced for

Fig. 7.7 A selection of small knobs.

the thread of the screw and with a recess for its head. Ideally, use hex-head screws – with a suitably shaped recess, these will work without the need for any adhesive. The underside of the knob should be bevelled so its outer edges are clear of the surface it is tightened against.

Sometimes, however, you will want a bit more security than a push fit, or you may only be able to get hold of cap screws. This is no problem as any high-strength adhesive can be used to retain the knob in position. For a particularly attractive result, use a coloured epoxy resin that matches the 3D-printed knob, and fill in over the head of the screw.

If three semi-circular 'scoops' are removed, a knob for higher torque is made, while many smaller grooves will give something more comfortable to use. Naturally, by adding bevel, fillets or other features, the knobs can be fine-tuned to meet your needs.

Producing these little knobs is almost addictive; you will be surprised just how many applications you will find for them.

3D-PRINTED EMERGENCY ER COLLETS

The ER collet system, which uses collets that are able to hold a range of sizes, rather than just suiting one size, have become very popular with hobby engineers. The vast majority of ER collets are designed to take round stock, although they can be pressed into service to take hexagonal and square bars. With 3D printing, an entirely new possibility arises (*see* Fig. 7.8). By designing a 'blank collet' and then 'subtracting' the shape of your workpiece, it becomes possible to hold virtually any suitably sized object in an ER chuck or collet block, even tapered parts, irregular objects or screw threads.

Clearly, plastic collets are not going to take the loads imposed by heavy

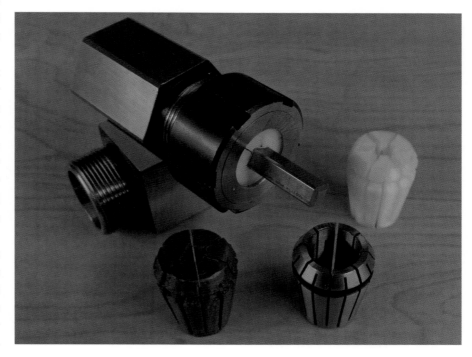

Fig. 7.8 ER25 collet block with 3D-printed emergency collets.

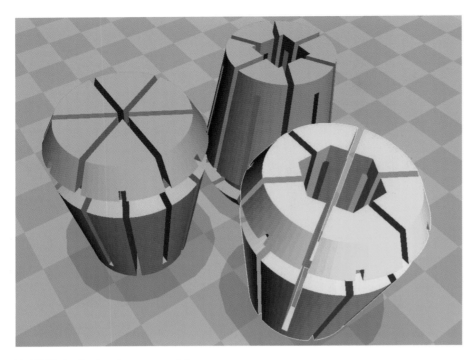

Fig. 7.9 3D visualization of emergency collets.

machining, but if cuts and feed rates are kept low, or the material is relatively easily worked, then this is a realistic option.

For best results, the example collets should be printed with their front face on the build surface, at a resolution of 0.15mm or better and at least 50 per cent fill. Use a raft to ensure that the front edges of the collet are nice and sharp.

8   A Simple Milling Cutter Holder

Anyone who has a milling machine should have a proper collet chuck for holding their cutters, such as the design in Chapter 9, as a good solid chuck is essential for the safe and accurate use of larger cutters. Nonetheless, there are times when it is a real advantage to have a smaller holder, such as for carrying out more delicate jobs or milling in a relatively small lathe. Even if you have only an occasional need to carry out such precision work, a good solution is to use 'throwaway' FC3 cutters. These are cutters made from high-speed steel for use by industry in vast quantities, and are therefore relatively inexpensive. They come in a wide range of small metric and imperial sizes, with either square or 'round' ends, and have three flutes allowing them to be plunged like a slot drill as well as used like an ordinary end mill. Although characterized as 'throwaway' – in an industrial setting they are simply recycled – they are not impossible to sharpen. If you have only a basic grinder, a cutter with chipped teeth can be rescued by grinding across the end at a shallow angle to leave one good tooth – just use it slowly afterwards.

These cutters have a shank size of 6mm for metric sizes and 1/4in for imperial. They are precision cutters and should be held in a holder that matches their accuracy. It is not difficult

Fig. 8.1 Milling steam passages.

Fig. 8.2 FC3 and other small cutters.

to make one of these from a suitable blank 'arbor'. Typically, you will need to use an MT3 arbor, which will fit most bench lathes and mills. Versions with easily machined 1in or 7/8in diameter sections on the end are readily available; in this case, choose a relatively small size.

Take more than usual care to achieve an accurately centred hole that is a close

Fig. 8.3 The simple MT2 cutter holder.

sliding fit for the cutters. The method is as follows:

1. Check the lathe is switched off.
2. Remove the chuck from the lathe and clean out the taper in the end of the mandrel with a tapered piece of wood followed by a piece of cloth on a stick, *not* a finger. The MT3 arbor should slip in securely, with a gentle twist to lock it in place.
3. Use a tailstock drill chuck to make an accurate hole in the end of the arbor.
4. Start with a well-lubricated spotting drill or a centre drill, followed by an intermediate drill to full depth. Approaching full size in steps helps ensure concentricity. Minimizing the extension of the tailstock barrel and 'almost locking' its movement both help with getting a good result.
5. Now drill several thou undersize, before reaming to the exact size. For 1/4in, letter C is ideal, or 15/64in at a push. If you prefer to use metric cutters, you will need to aim for a final size of 6mm, so a 5.8mm or 5.9mm is ideal. Reaming needs

care. Look for good tailstock alignment, the right amount of material to remove, speed rather slower than for drilling, and some cutting fluid, applied by brush or oil can – you do not need a continuous flow.

6. Make the cut with steady pressure.

If you do not have a suitable reamer, you can make a D-bit from a piece of 1/4in silver steel (*see* Chapter 5). In use, a D-bit works like a slow drill. Keep it well lubricated and feed it in slowly, and

it will produce a remarkably accurate hole.

To lock the cutters in place, a radial hole for a set screw is needed (*see* Fig. 8.4). You could use 2BA or M5. Drill the hole on the drill press with the arbor held in a vice. A centre pop will help ensure the drill does not 'skid' on the surface. This screw engages with a flat on the shank of the mill.

Milling cutters tend to 'grab' more readily than drills, pulling the taper from its socket, with disastrous results for the work and for the cutter. It is essential to use a drawbar – a rod screwed into the end of the holder, tightened against a stop at the far end of the mandrel. MT3 arbors are typically threaded M10, but 3/8in BSW is still regularly encountered, so check the thread and do not assume it is M10. The drawbar used here has a knurled nut and a separate plug to fit the end of the lathe spindle (Fig. 8.5). If you do not already have a suitable drawbar you could use a length of threaded rod with a suitable washer and nut, or make your own drawbar, threading the ends of a rod in the lathe.

Fig. 8.4 Grub-screw fixing aligns with the flat on cutter shank.

Fig. 8.5 Drawbar fitted to cutter holder.

Fig. 8.6 Drawbar disassembled.

9　An ER25 Collet Chuck

MAKING A COLLET HOLDER

ER25 collets are probably the most ubiquitous choice for holding milling cutters today, offering high levels of grip and available with varying degrees of accuracy. Even the standard ones are good enough for ordinary milling work. They can also be used to hold cylindrical workpieces. Many years ago, I made myself an ER25 collet holder. At the time it appeared that it was only possible to get blank MT arbors with relatively small soft ends, so for my design I was forced to use a fairly hefty chunk of steel screwed on to the end of such an arbor (Fig. 9.2). It also used a much larger closing nut than is usual. The resulting chuck had a couple of advantages: I was able to drill out the end of the arbor, so relatively long workpieces could be inserted right into a collet; and the sheer mass of the chuck gave it extra momentum, which helped on interrupted cuts when using it to hold a fly cutter. That said, it was rather clumsy, and its length used up too much headroom in the mill, although it works as well today as when it was made.

More recently, MT3 blank arbors with large-sized ends have become available – large enough to form the body of a chuck in one piece with the shank. In addition, inexpensive ball-bearing closing nuts offer a compact and effec-

Fig. 9.1 ER25 collets.

Fig. 9.2 Original design ER25 chuck.

Fig. 9.3 Original chuck in use.

tive way of closing the chuck, while eliminating some of the rather challenging machining that would otherwise be needed to make the extraction ring.

The arbor used had a blank end 1 1/2in diameter by 1 1/2in long (38 × 38mm), which was big enough for a one-piece collet holder. Clean out the mandrel taper of the lathe (to help ensure accuracy) and fit the blank, securing it firmly in place with a drawbar. Turn down all but a 10mm collar at the base of the blank to 32mm diameter, and then put a runout groove 1.6mm deep and about 6mm wide in between the collar and the 32mm section, using a parting or grooving tool. Using suitable change wheels, cut a 1.5mm thread on the 32mm section, using the new nut to check the fit. The blank was not too soft, but easily turned to give a lovely bright finish.

Drill the centre of the blank out to 13mm (or the largest drill you have up to this size), in preparation for boring the taper.

I still have the small 82-degree/8in square I made at the same time as my first ER collet chuck (Fig. 9.5), so I used that to set the top slide at the correct angle. Otherwise, use an engineer's protractor or 'eyeball' the angle from a collet. I then bored out the blank until an ER25 collet would fit inside, with

Fig. 9.4 ER25 closing nut fitted to collet.

Fig. 9.6 Fine finish inside taper socket.

Fig. 9.8 Original and revised ER25 chucks compared.

the release groove about 4mm from the face of the blank. Depending on how accurately you have set the angle, you will probably need to make some adjustments. Start testing the angle well before you have opened the bore right up. The best way is to very lightly mark the cavity with engineer's blue and gently rotate a new collet in the bore. If it only removes blue from one end of the bore, adjust the angle a tiny amount and take a very fine cut.

If you are not confident in setting angles this way, practise by boring a piece of scrap (held in a three-jaw chuck) until the angle is spot on, then swap the chuck blank for the practice piece without changing the taper setting.

A final test using a collet and the nut should show that everything works as expected. You can give a final polish to

the collet bore by cutting a piece of very fine emery tape to fit around the collet without any overlap. Smear it with clean cutting oil and gently hold the collet and emery in the mouth of the rotating chuck. The final finish should be a dull but very smooth matt finish.

The final task is to drill three 6mm holes, evenly spaced, around the 'collar' for a tommy bar or C-spanner (Fig. 9.7). There are various ways of accomplishing this; one simple option is to cut a piece of paper to fit around the circumference, fold it in three equal parts, and then use it as a guide to scribe three marks. Centre pop and drill at each mark to a depth of just 9mm, which will avoid the risk of breaking into the bore. A 6mm diameter tommy bar is best made of silver steel, hardened and tempered to a straw colour, as a mild-steel

bar will bend under the force needed to close an ER collet.

Fig. 9.8 compares the original ER25 collet chuck with the much more compact version using a one-piece arbor and commercial closing nut. The ball-bearing nut makes this collet holder fast and pleasurable to use. As for the accuracy of this arrangement, I put a dial indicator on the shank of a random end mill fitted to a random collet without any special care. There was barely visible runout – about a fifth of a 0.0005in division, or a tenth of a thou. That is about 5 microns, which is about the normal tolerance for standard ER25 collets, suggesting that turning and screwcutting the holder in place gives a perfectly acceptable result.

DRAWBARS

Making a Self-Releasing Drawbar

The boring head should always be used with a drawbar to keep it securely in place. This self-releasing drawbar design greatly speeds up changing tooling and is intended to fit the popular X2-style milling machines, but it can be readily adapted to many other benchtop mills.

Fig. 9.5 82/8 square made to aid turning ER tapers.

Fig. 9.7 Finished ER25 chuck.

MAKING A TOMMY BAR

If you choose to use a tommy bar with your chuck, you will need to make one. It may seem that there is nothing simpler, but if a properly fitting tommy bar is not to hand, it can be tempting to use anything of approximately the right size. This brings a number of hazards: the wrong size or shape (perhaps using an Allen key) can damage or bruise the hole, while a bar that is too long can impose undue stress. A handy length of brass or mild steel might do the job just right, then be returned to the scrap bin with a bend that does not become apparent until it is needed for some other job. It is far better to have a well-fitting, appropriately sized tommy bar handy to cover every job. You probably do not need instructions on how to make it, but do take some time to finish it nicely, so you remember to keep it separate and out of the scrap bin. For most steels, blacking in oil will do a good job.

Fig. 9.9 shows a bar I made to suit a number of tools, including a collet

Fig. 9.9 Tommy bar.

chuck. One end is slightly reduced so each end fits different sizes of hole. A carbon-steel bar can be heat coloured or blacked, but this was made of stainless steel so it is quite strong, but impossible to black. Instead, I originally wrapped it carefully with rings of insulating tape. While this lasted a couple of years, it eventually started to peel off. I therefore replaced this with heat-shrink tubing as used in the electrical industry. Just thread a length over the centre of the bar and warm it with a hot-air gun, or a small gas lighter flame from a suitably cautious distance.

On some occasions – for example, when using small vices or clamps – a captive tommy bar is a suitable

solution. Tidy upsetting (forging to greater diameter) of the ends of such bars is difficult and simply flattening the ends (as is often done on cheap disposable G-clamps) looks cheap. A neater trick is to make the bar a good fit in its hole and use a centre punch to raise two hollow 'craters' at each end of the bar. These are surprisingly effective at stopping a well-fitting bar from slipping out and make an excellent solution for small clamps. A deluxe solution is a ring of metal held in place by a force fit at each end of the bar. This is a very neat way to put ends on a bar, it gives excellent practice in turning the right fit. To make it easier, slightly taper the ends of the bar, drive them into the rings, then file the ends flush.

This may seem to be a lot to say about a simple metal bar, but it is worth remembering the advice of the late George Thomas, who always advocated taking the greatest care with even the simplest of jobs. That way, you will acquire the skills to undertake really first-class work.

Most benchtop milling machines are supplied with basic drawbars, which are simply rods threaded at each end and fitted with a nut and washer that are tightened on to the end of the spindle. These drawbars work well enough at keeping tapers from working loose but offer no help in releasing the taper. You typically end up choosing between tapping the end of the drawbar with a hammer or poking a brass bar down on to the end of the taper and tapping that. Either way, the taper and

attached tooling eventually drops out and rolls off the bit of wood that you

Fig. 9.10 Original drawbar arrangement.

(hopefully) remembered to position underneath it.

A self-releasing drawbar, suited to any machines with a threaded top to the spindle, is easy to make and renders the removal of tapers a simple and stress-free exercise. Instead of the top nut and washer, the drawbar has a thick collar (Fig. 9.11) attached to it, which is held captive by a top cap on the end of the spindle (Figs 9.12 and 9.13). It tightens up by means of a square on the end, and is released simply by unscrewing this square so that the collar presses against the inside of the top cap and then the thread gently forces the taper out. If you use one hand to hold the tooling (perhaps with a tommy bar) and the other to turn a spanner on the squared end of the drawbar, the tooling simply pops loose, then drops into your hand as you spin the drawbar right out.

As there are many different small mills, the instructions for this design

Fig. 9.11 Self-ejecting drawbar.

Fig. 9.12 Top cap for self-ejecting drawbar.

Fig. 9.13 Top cap fitted.

should be taken as a guide only, not as definitive. Check the sizes of your own machine. You can remove metal from the bottom of the collar to get more engagement of the drawbar in the end of the taper, and from the bottom of the end cap if the drawbar is too short to release the taper.

You will probably need to make two drawbars. Typically, the threads needed are 3/8in BSW and M12, though some tapers have an M10 thread. As long as your top cap has a 1/2in hole, both metric and imperial drawbars will fit and you will be ready for anything!

Ideally, use a tougher steel such as EN24T or EN18T to ensure long life, although a plain mild-steel drawbar will do the job as long as it is not over-tightened. It is a good idea to make the drawbar blank over-length and then silver-solder an oversize collar in place. You could use high-strength retainer and a cross pin at least 1/8in (3.5mm) in diameter. You can then turn the collar to size and cut the bar to length before making the thread. A die-cut thread should be perfectly adequate, but, as the top cap requires you to set up for thread cutting anyway, you could use the same pitch for both threads.

The square on the drawbar can be milled or filed. The square allows you to tighten it up with a spanner, but keeping it small (even on an M12 drawbar) makes it less likely you will overtighten it. The drawbar is there to make sure the taper is fully engaged and does not shake loose, not to forcibly hold the taper in place. The MT3 taper is a self-holding taper and over-tightening will just make releasing it more difficult.

Fig. 9.14 Spanner square on end of drawbar.

Modifications

Two minor modifications are needed to the mill. The most significant is to thin down the top spindle bearing locking washer, to provide enough free thread for the top cap to lock in place (Fig. 9.15). The second modification is simply to drill a hole in any plastic spindle top cap, to allow the new extended drawbar to poke through it.

Top Cap

This is the biggest task, because you have to hollow out the top cap. Obtain some free-cutting mild steel of the right outer diameter and starting by drilling the hole through the centre.

Fig. 9.15 Position of collar.

Fig. 9.16 In use with cover fitted.

Fig. 9.17 Left-handed thread inside top cap.

The thread in the top cap has to fit the top end of the spindle. On this X2 mill it is left-handed, 16tpi (Fig. 9.17). Metric aficionados should find a 1.6mm pitch thread will work in this situation, the thread is not heavily stressed and only engages a few turns. You should find no special difficulty in cutting the left-hand thread, as it is a relaxing change to have the cutting tool emerging from the work, rather than disappearing into unknown depths. Just make sure you have set up to cut a left-handed thread!

The 6mm (1/4in) hole in the side of the cap is important. If the drawbar is in place for an extended period, the top cap can tighten it up a little (it only needs to be fitted finger tight). If the drawbar does not get swapped over very often, the cap can get very tight. The hole allows a C-spanner to be used to get the top cap loose. That supplied with the mill for adjusting the spindle bearings does the job nicely. This is much easier than making a couple of flats for a large spanner. Incidentally, you may not have noticed that there is a small hole in the right-hand side of the top plate of the milling head. You can insert a 6mm bar in this hole to lock the spindle – just do not forget to take the locking bar out before starting the mill!

Finally, now you have made a nice job of this simple accessory, it really deserves a good finish. Oil blacking the cap, while leaving the drawbar itself bright, makes a pleasing contrast. If you do decide to carry out this modification, you will not regret it, as it will make any tooling change quicker, easier and less stressful.

10   A Small Dividing Head

On many bench lathes there is little space for even the smallest of rotary tables on the cross slide, particularly with a chuck attached. The ML7 can benefit from an extra-long cross slide, but this option is not available for many other lathes, and even with the extra space for attaching something a full-size dividing head is heavy and bulky.

The small dividing head in Fig. 10.1 was made entirely from odds and ends, aside from its chuck. As the smallest rotary tables can be awkward to use on smaller lathes, and if your lathe does not take screw-on chucks, the George Thomas type of dividing head can be difficult to adapt without making exces-

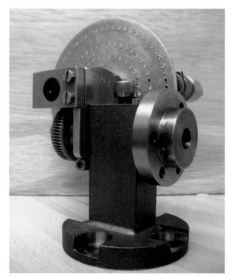

Fig. 10.1 The small dividing head.

sive overhang or a custom backplate. If you have these problems, the advice below will help you to use a small, inexpensive lever scroll chuck to provide a device for precision division work on small lathes.

Taking inspiration from an un-dimensioned plan in a mid-1940s issue of *Model Engineer*, I made a small dividing attachment that appeared to offer a solution. Sadly, several parts of my original light dividing heads have since been scrapped, but I have reassembled enough of them to give an indication of how they once looked. Also shown are two cotter pins, a spare spindle, some taper mandrels (all but one are unused blanks) and the spindle used to gash the gear blanks and drill the index plates (Fig. 10.2). The drawing was a reprint from an article by Edgar T. Westbury from the inter-war period. I followed the design reasonably closely: a simple head and tailstock, mounted on a bar held in a toolpost-mounted body (Fig. 10.3). The whole device rather resembled a watchmaker's turns, and could serve as such, although the provision of a geared spindle in the headstock makes it far more versatile.

The body and end stocks were made from mild-steel bar. The critical part of the operation was making the bores of the two end stocks truly parallel and equally spaced. This was achieved by

Fig. 10.2 Parts of the Westbury dividing attachment.

Fig. 10.3 Old-style tailstock.

soft-soldering the two pieces together and relying on holding the pair firmly against the body of the four-jaw chuck. In the end I produced a rather simpler shape than Westbury's elegant stocks, which were presumably made from castings.

The main bar was a length of 3/8in stainless steel and the spindles were made from a large high-tensile bolt. The most involved tasks were making the internal taper on the nose and milling the keyway. The latter job was done with the spindle locked in the headstock, itself clamped in the toolpost.

Instead of boring the spindle for 8mm collets, I decided to bore a simple self-holding taper and provide a small drawbar. The taper was made with a simple silver-steel reamer, turned to size and ground halfway through before hardening and tempering. This was run into a pilot hole with plenty of oil, without any problems with chatter. I turned up a handful of matching blank tapers in mild steel at the same time, which proved to be a wise move. The drawbar was from 1/16in silver steel and can be seen fitted to a hardened 60-degree centre alongside two spindles in Fig 10.4.

The dividing part gave me the opportunity to use the attachment to cut its own worm wheel by free hobbing. I later used it to produce a larger wheel for a rotary table. Both feats were accomplished in the same way, following a simple technique outlined in *M.E.* by Geometer in his 'Workshop Tips' during the 1960s. The first step was to turn a flat-topped thread on silver-steel bar. The worm for the reduction gearing was turned to 20tpi. This was chosen as I had a length of 1in by 1/4in brass flat for the wheel, and this pitch was easy to set up and gave as near a 1in wheel as possible. One portion of the thread was left a few thou 'full' and this was separated, fluted with an end

Fig. 10.4 Spindles, taper centre and drawbar.

Fig. 10.5 20tpi hob for cutting worm wheel.

mill, hardened and tempered for use as a hob (Fig. 10.5). The other portion was further machined to form a suitable spindle. Before leaving the lathe, I also turned up the blank for the gashing tool, which was finished by filing. Both hob and gashing tool were crudely relieved using a grindstone in a mini-tool on the outer edges only.

Turning up a blank was no problem, starting with a 25mm (1in) square bored and keyed to a special spindle for the dividing head, but held in the three-jaw chuck. This was then transferred to the headstock of the dividing attachment. A 60-tooth change wheel was fixed to a short mandrel held in the taper at the opposite end of the spindle, as shown in the photograph. Indexing was facilitated by the improvisation of a simple detent of spring steel, which allowed the change wheel to be indexed around one tooth at a time. Each tooth-space was carefully gashed out with the above-mentioned tool, held in the three-jaw chuck. The attachment showed its worth at this stage, as it made angling the blank to get the correct gash angle very easy – a job that was done by eye. Gashing took two attempts – the first failed, as a slight movement of the main shaft meant that cut 61 was not in exactly the same place as cut 1. The second time, I took extra care to clamp everything firmly in place,

but the difficulty in clamping the bar continued to be a problem.

The next stage proved straightforward – the spindle was set so that the blank was horizontal and at centre height, and the hob was mounted in the three-jaw chuck. With more experience I should have used the four-jaw, but the chuck was in good condition, and the hob ran true enough. Helped by the free-cutting brass, spinning the hob at about 200rpm and gently advancing the cross slide, the wheel cut itself. Every so often, I gently moved the saddle from side to side a little, so that more than one part of the hob did the cutting. I finished up the job by running the wheel along the full length of the hob. To cut the rest of this part of the story short, the next task was to fabricate the worm-cage from brass flat, relying on hand fitting to get a good mesh with the worm wheel. In short, I was very pleased with the quality of mesh.

I used the same technique to make the worm and wheel for a rotary table. This was to be 2in diameter, to match a 10tpi worm. My initial attempts were a

Fig. 10.6 20tpi worm and wheel.

Fig. 10.7 20 and 10tpi hobs compared.

fiasco, until I realized that I was wasting my time using a sheet of CZ108, a hard grade of brass. While this could make excellent pinions if cut by a proper tool, a basic home-made hob literally bounced off it, unless I used so much force the set-up distorted. I ordered a piece of leaded CZ120 'engraving' brass and got a decent result first time. The remnant of CZ108 eventually became a gib strip for another project.

I had left the end of its spindle rather short. Once I had made a mounting bracket for the index plates, it only had a short stub on which to mount a simple handle and detent, and there was no space for attaching index arms. The original detent, shown alongside an index plate simply screwed in and out (Fig. 10.8), was replaced with a sprung detent.

Making the index plates was thera-peutic – although a few hundred holes is not in the same league as several thousand tender rivets! I wish I had

taken care to space the rows neatly. They were meant as an experiment and ended up good enough to use. The plates are just 16-gauge aluminium, so I do not expect them to last for ever, but I will not replace them until they start to wear, at which point I will make new plates from brass. The design of the blanks allows them to fit directly to the end of the spindle in place of the worm and wheel for direct dividing. By using change wheels on the other end of the spindle, in the same way as for gashing the worm wheel, indexing the index plates was easy. Each hole was centre-drilled to the same depth in the lathe; a high speed was needed, but the process resulted in neat conical holes.

The handle itself was made up as it went along (Fig. 10.9). While the detent is a bit clumsy, the free-turning brass handle is pleasant to use.

Despite the success in making the head and tailstocks, I was frustrated by the difficulty of gripping the 3/8in rod at the heart of the device. It appears that the central spindle in Westbury's design was split, so that it gripped the rod as it was drawn in. Despite many tweaks, I could not get this arrange-ment to give adequate grip for even the lightest operations, and I wonder how

Westbury got it to work. A pin running in a slot in the bar met with some success, but it lacked accuracy and weakened the clamp, which then broke. The rod could be fixed in the 'holder' allowing the whole assembly to rotate but not move lengthways. In contrast, Westbury's cotter pin arrangement for the end stocks worked perfectly.

In any case, the device was suitable only for very light tasks. Workhold-ing was not simple either, essentially limited to small mandrels made to fit the custom taper socket or holding work between centres. A simple face-plate could have easily been made to fit the device, but a small chuck would have been an ideal solution.

At the other end of the scale, the rotary table I had made in the meantime was too big with its solid 4in square base. It was fine when I finally got a milling machine, but you could see the worktable droop when it was mounted on the drill press. With a chuck attached there was hardly any space left to use it on the lathe or with large drills.

Some sort of compromise was needed – a device that was smaller but still had reasonable workholding capac-ity and good rigidity. I realized that its small scroll chuck could be attached to the dividing mechanism from the West-bury device. I obtained a 63mm scroll chuck, choosing a plain back rather than a 14mm Unimat mounting. This was meant to be mounted on a spigot and secured with three screws.

To give a solid base, a slice was cut from 2in diameter mild steel and trued up in the lathe. A slot was milled to take a section of 7/8 × 5/8in bar as an upright. Four recessed mounting

Fig. 10.8 Index plate and detent.

Fig. 10.9 Handle and index plate fitted.

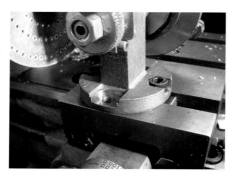

Fig. 10.10 Method of attachment to lathe.

holes to take 6mm mounting bolts were milled in the base to allow T-slot mounting along and across the axis of the lathe (Fig. 10.10). The front recess was needed to ensure the bolt did not interfere with the chuck; the others allowed all the bolts to be the same length. Two counterbored 5mm cap screws held the upright in place on the

base – careful hand fitting was needed to get it vertical.

With the two parts firmly joined, the base was mounted on the lathe, and a dial indicator in the chuck used to check it was square when the saddle was traversed. With a centre in the headstock, the base was lined up crosswise, and the saddle locked by tightening all three gib screws. A 1/2in hole was then bored and reamed to size at lathe spindle height. Fig. 10.11 shows the overall arrangement.

A previously made brass bush was fitted in the hole, and itself bored just less than 3/8in at the same setting. One end of this bush had a 1/8in by 0.550in diameter flange, to mount the dividing apparatus, and a 7/16in ring made to fit the other end. These parts were well

fluxed and assembled to the upright, and silver-soldered in place.

The job of cleaning these parts was an opportunity to try out a sulphamic acid-based limescale remover as a pickle. Used while the work was warm (not hot), it removed the easyflo flux in a few seconds, leaving the surface very clean. A Dural bar was gently wrung into the bore as a mandrel, and skimming cuts taken to clean up the two bushes. Finally, a 3/8in reamer was put through the bush by hand.

The chuck needed a backplate of 36mm in diameter, but the spindle would be only 3/8in in diameter. I did not want to turn the entire thing out of a single chunk of steel and, as luck would have it, that day I found an M16 bolt big enough to make the spigot, but not the flange, out of. Holding the head of the bolt in the three-jaw chuck, I carefully faced and centred the other end. Supporting the small end with a half-centre, I roughed it down to just over 3/8in. I also turned a true face on the inside of the bolthead. I then reversed the bolt and turned the head down to just larger than the chuck's 22mm register. I drilled a blind 1/4in hole and, using a home-made reamer, opened this into a taper to take the stock of little mandrels made for the original dividing head.

I then drilled a 3/8in hole in the middle of a piece of 1/4in steel plate. This was opened with a file until it was an easy fit on the spindle, and silver-soldered to the back of the bolthead. Rather than finish turning the spindle between centres, I continued to use the three-jaw chuck for workholding. With the narrow end of the spindle in the three-jaw chuck I found I could get

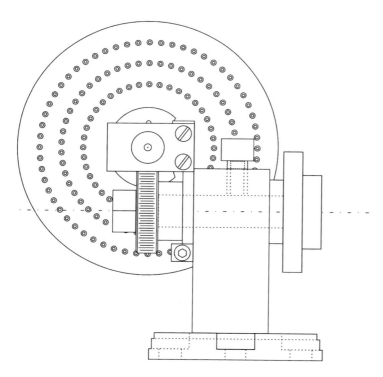

Fig. 10.11 General arrangement of small dividing head.

Simple Dividing Head - General Arrangement

concentricity of no worse than a thou, and I went ahead and turned the back-plate and the register to diameter (but left this over-length at about 9mm).

I cleaned the taper socket and popped in a taper mandrel with a 3/4in long stub on the end. I had produced a collection of these at the same setting as I used to make the taper reamer. The spindle now ran true, so I supported it with the tailstock centre and started work. Aside from the section that has to be a good sliding fit in the upright, I copied all the dimensions of the rest of the spindle from the shorter spindle of the light dividing attachment. This meant I could use the original spacer, worm wheel and fixing nut (Fig. 10.12). Finally, I drilled through 3/32in and carefully tidied up the end of the spindle without the tailstock centre in place. The trusty little mandrel was removed, and I then reversed the spindle a last time, reduced the length of the register to a shade under 6mm, and re-bored the taper to its final depth. The three M4 holes for the chuck-retaining screws were positioned by spotting through from the chuck itself. I hand filed a flat for the spacer grub screw and milled a 1/16in slot for the worm-wheel key. The latter operation was performed with

Fig. 10.12 View of spindle.

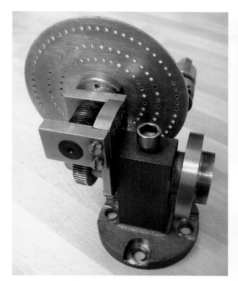

Fig. 10.13 Spindle locking arrangement.

the new spindle locked in the body of the original light dividing head!

Provision for locking the spindle was added with a simple M6 tapped hole in the top of the pillar (Fig 10.13). A concave brass pad shaped to a good fit on a 3/8in bar was prepared to drop in the hole and put to one side with a bright plated screw.

The device was reassembled, to ensure that everything lined up. It was then broken down, and the two parts of the base were separated and degreased. They were warmed over the workshop heater and reassembled with a little two-pack epoxy in the joint. Small fillets of epoxy were made around the join to help create the illusion of a casting. The base was given two coats of matt-textured Hammerite before final assembly.

The result was a small but versatile device. On the lathe it gives an excellent set-up for delicate rotary milling operations, such as milling ports in the cylinder barrel of a small compression

Fig. 10.14 Using the dividing head to mill engine ports in the lathe.

Fig. 10.15 Using the head to drill holes in a flange.

Fig. 10.16 Milling flutes.

ignition engine (Fig. 10.14) and drilling holes on a diameter. This was a delicate operation that would have been clumsy to do on even a small milling machine, but it was easy on the lathe. Despite the relatively small size, the device is rigid enough for work of a size that can be held in its small lever scroll chuck and is certainly suitable for light milling operations.

11   Making a Fly Cutter

A fly cutter is a single-point tool, usually used for machining plane surfaces. It is an inherently accurate tool when used in this way, and if the toolbit is a good one, rigidly held, it will produce smooth, flat surfaces with the most basic of set-ups. It is also possible to fly cut convex surfaces, but the lack of fine adjustment of the typical fly cutter imposes limits on accuracy.

There are many ways of making a fly cutter. The basic requirement is simply a toolbit holder, which is easy to attach to a rotary machine such as a lathe or a mill. This straightforward but useful example was inspired by a photo of one made by John Rinaldi – many similar ones are manufactured. Stan Bray has described another useful style, which bolts directly to the faceplate. Tubal Cain was not ashamed to use a square-section boring bar held crosswise in a chuck!

Fig. 11.2 The body of the fly cutter, showing the slot.

The body of this fly cutter is just a piece of 25mm (1in) mild-steel bar, turned down to 1/2in to make a shank (Fig. 11.2). When you hold the cutter in a chuck, the body can be held flush against the jaws, preventing any rear-ward movement. The thick end of the bar can be sawn and then filed to a 10-degree angle. This angle gives the toolbit its top relief.

For use with 1/4in high-speed steel (HSS), a slot offset from the centre line of the body is needed. On my lathe I was able to achieve this by holding the shank in the lathe's toolholder, suitably angled. I cut the slot with a 1/4in FC3 mini mill, which gave a slot that was a good fit for the tool steel. By using the same mill in a pillar drill and 'plunging' it into the thick side of the body, two recesses with 4BA tapped holes for the securing screws were easily made.

The tricky bit is making a good toolbit. You will need a piece of 1/4in square HSS roughly 50mm (2in) long. If you need to shorten a longer piece, grind a shallow groove all round with a mini drill and carborundum wheel. Hold the bar in a vice, covered with a cloth, then tap the end sharply with a hammer. To visualize the shape of the tool, think how it will contact the work as it rotates; with the bit in the holder, identify the corner

Fig. 11.1 Fly cutter with HSS toolbit.

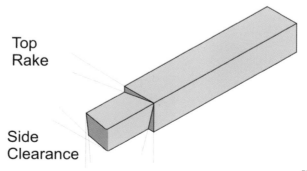

Top Rake

Side Clearance

Front Clearance

Fig. 11.3 A suitable toolbit for general use.

that will do the cutting. The angle of the holder provides one clearance angle, but you need to grind further clearance on the end, 'front' and 'top' of the bar. Aim for 5–10 degrees, and a little more on the end of the bar. Effectively, you need to create a left-handed knife tool.

In use, the cutter should be held firmly in a chuck (*see* Fig. 11.4). In the case of milling machines supplied with heavy-duty drill chucks, you can probably get away with holding the cutter in it, but do not try using the fly cutter in a pillar drill. Fly cutters need to be run slowly;

Fig. 11.4 The fly cutter in use.

this one has a maximum of a 3in cut and you should run it no faster than you would turn a 3in diameter cylinder. The work should be moved past the cutter. On smaller lathes, it can be a challenge to achieve a set-up where the whole area to be worked passes the cutter. The depth of cut and feed rate should be less than usual, because of the significant leverage between the end of the cutter and the shank. Be patient and keep your cutters sharp and you will get excellent results.

Finally, a version of this tool that is not angled can be used to hold form tools, such as those used for cutting gear teeth or special profiles.

12　Boring Tools

Boring tools and boring bars are used for making accurate round holes. They may be used in lathes or milling machines and the basic requirement is a single-point cutting tip on a stiff, overhung bar that can fit inside and reach through, or to the bottom of the required hole. The great majority of such tools are not suitable for making holes from scratch, but rather for opening up an already drilled or cast-in hole. Although this requirement is simple, boring bars come in a multitude of designs: some are ground from a single piece of HSS (or, in the old days, carbon steel), while others take inserted cutters of various styles. Some, but not all, of the latter type are adjustable.

As a general principle, always choose the shortest, most robust boring bar for any job – the stiffer it is, the better the surface finish will be and the faster you will be able to remove material. Assuming you have a stiff enough tool, depth of cut, cutting speeds and feed rates are the same as for normal turning.

Fig. 12.1 A round-bodied, double-ended boring tool with a split block holder.

However, if you are working with a long or narrow boring bar, you may have to drastically reduce depth of cut and feed rate.

Milling machines typically use single-piece boring bars held in a boring head that allows them to be adjusted for cut diameter (*see* Chapter 13 on making your own boring head). The work is fixed to the mill's table and adjusted for position, then repeated passes of the tool are made until the required diameter is achieved.

For work being turned on a lathe and requiring a hole to be opened up, the usual solution is to use a boring tool held in the lathe's toolpost. The resulting hole will be concentric with the lathe axis and any other turning done at the same setting, and sizing becomes a straightforward case of adjusting the cross slide to put on the cut. Of course, work can be mounted on the faceplate or an independent jaw chuck if the required hole is not concentric with an outer diameter.

For large work, options for the lathe are either a between-centres boring bar, threaded through an existing hole in the work, or a single-ended or cantilever bar that may be held in a four-jaw chuck. This requires the work to be set up accurately on the cross slide. (A T-slotted or drilled and tapped plate fitted on the cross slide to mount such

work is referred to as a 'boring table'.) This process can be particularly fiddly. Holding the work in a vertical slide does much to facilitate accurate positioning of the hole.

A between-centres boring bar is stout bar driven from the headstock and supported at the tailstock end, with an adjustable cutter at its mid-point. This is a very rigid solution that gives excellent results, more or less guaranteeing a high degree of roundness and parallelism in the bore. Some larger between-centres boring bars use a graduated screw to advance the tool to size the hole. On smaller bars, a micrometer across the tool tip and the main bar can be used to measure increases in the depth of cut.

Adjusting the depth of cut by offsetting the tool in the four-jaw chuck can be difficult to do accurately but can be a solution where the size of the hole is not critical. It is also possible to mount a boring head in the lathe spindle, in which case the exercise is essentially the same as boring in a milling machine.

GROUND BORING BARS

There are two styles of fully ground boring bars. Those primarily intended for use in the lathe typically have a square shank to fit in the toolpost, a reduced section (to allow chips to clear easily) and then what is effectively a small knife tool ground on the end. Often, they will be used with a holder that places the cutting edge at centre height (*see* Fig. 12.2); otherwise, shims may be used under the tool shank. These tools generally will work in holes with a diameter significantly larger than the size of the tool tip; it is often possible to modify them for smaller holes by grinding away below the cutting edge. This will give a curved profile without significantly reducing the tool's effectiveness.

Most often, these tools are ground from high-speed steel. Boring bars with brazed-on tungsten carbide tips are available but are often (although not always) rather crude. The TCT-tipped ones can be a godsend if you have to

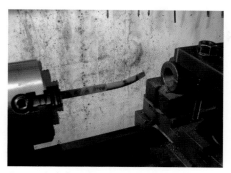

Fig. 12.4 Carbide tool mounted in lathe chuck in order to bore a tough casting.

bore out cored holes in a casting with a particularly hard 'skin' as in Fig. 12.4.

The other style, used mostly with milling machines with a boring head, is a round-shanked 'cousin' of the D-bit, which it strongly resembles, although it usually has a reduced portion for chip clearance (Fig. 12.5). It can be held in a lathe toolpost using a split bar holder or special toolholder blocks. These tools have round cutting ends and will work in holes only marginally larger than the diameter of the tip. It is also possible to press D-bits into service as boring bars, and to use boring bars as D-bits.

If a long, thin boring tool is needed, it is possible to produce a very effective makeshift one with maximum rigidity

by angling the end and grinding away slightly more than half the diameter of the tip of a length of round HSS. This is then mounted in a holder so it is slightly angled to give clearance to the cutting tip and rotated so the cutting point is dead on centre height. Having 6mm and 3mm tools of this type handy can save plenty of time; they are easily mounted by fitting a square block in the lathe toolpost and drilling and reaming it to size; a couple of set screws will keep the bars in position.

INSERTED CUTTER BORING BARS

There are many variations of inserted cutter boring bars. Naturally, such designs are only available in larger sizes.

A typical design for lathe use is a cylindrical bar with a clip or hole and clamping bolt for a short length of high-speed steel (Fig. 12.6). These are typically held in a suitable block so they can be angled to get the best cut. Sometimes they are double-ended, with one end holding the tool perpendicular to the shaft and the other at 45 degrees – used to reach the bottom of a blind hole (*see* Fig. 12.7). One benefit of this type is that square or round HSS in suitable sizes is readily available, and you can easily grind new or even special-purpose toolbits. Examples of these might include screwcutting tips for internal threads or form tools to make square or round grooves for rings or seals to be fitted within a bore.

Increasingly, however, tipped boring tools take standard tungsten carbide inserts. Those for lathe use are more likely to have square shanks, or at least round shanks with flats top and bottom,

Fig. 12.2 Square-bodied boring tool in a holder.

Fig. 12.3 Boring tool ground away beneath the cutting edge to improve access to small holes.

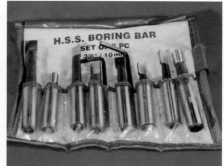

Fig. 12.5 A set of round-bodied one-piece boring bars, suitable for use with a boring head.

Fig. 12.6 Inserted HSS tip boring bar in quick-change toolholder.

Fig. 12.7 Using angled inserted tip to cut a square-bottomed blind hole.

Fig. 12.8 Boring bar with tungsten carbide insert. The body has flats so can be used in a holder or directly in the toolpost.

as the inserts have precisely set cutting angles, while those for milling have round shanks. It is possible to get insert cutter boring tools with built-in diameter adjustment combining the benefits of a boring head and inserted tooling.

HOLDERS FOR ROUND BORING BARS

The simplest holder is just a piece of square bar with a hole (to suit the boring bar) in it and split along one side. Clamping this in the toolpost squeezes the holder and holds the bar in place. You can be a bit cleverer than the suppliers of these devices by eliminating the need to pack up such a holder. Using a flat surface, work out how high the tip of the boring tool is when the tool is correctly oriented. It will usually be a small amount higher than the centre of the bar, so measure this distance by subtracting half the diameter of the bar from the height of the cutting tip. Now fit a suitable piece of steel bar in the toolpost, packed up by this distance and, using drills and then, ideally, a reamer driven from the headstock of the lathe, make a hole to closely fit the boring bar. Slit one side of the holder. It should now, without the packing, hold the cutting tip exactly on centre height. If the holder checks out, tidy it up and black it by heating to a dull red and plunging into clean oil.

Various other designs of boring toolholder can be used, such as that in Fig.12.9, and this technique can be used to ensure any of them hold the tool accurately on centre height without packing.

Fig. 12.9 Quick-change toolholder with shim for different-sized bars.

Although the simple split bar holder will serve for most applications, there are alternatives. If the boring bar shanks are too large to be held directly in the toolpost in this way, then a shanked holder can be made. The body can be split half way as before but will need to be wider to take one or two clamping screws to close it up. If the material to hand does not allow this, the holder can be taller and split vertically or have clamping screws that bear directly on the boring bar. If you do not have a large enough piece of bar to make such a holder in one piece (or don't fancy chewing a holder out of a larger block), you can silver-solder the shank to the body.

Smaller-shanked holders can also be very useful for holding very small boring bars when access is limited. However, a small bar in a small holder is prone to chatter and deflection. In extreme cases, a small boring bar may chatter so much that an ordinary twist drill may give better accuracy and surface finish. If this happens, the ideal solution is to drill undersize and use a reamer or D-bit to finish the hole.

13 A Micrometer Boring Head

This boring head is the most advanced tool in this book. It demands a certain amount of care in construction but you will be rewarded with a handy tool that should last a lifetime. The drawbar detailed in Chapter 9 is ideal for use with it.

A boring head is a tool that mounts in the bore of a lathe, mill or boring machine, and holds a suitable boring tool in an adjustable slide. Such an arrangement allows very accurate holes to be made to a high degree of precision across a wide range of diameters. The use of a graduated screw to adjust the tool allows very small adjustments to the cutting radius to be made. The boring head should be used with 10mm or 3/8in shank HSS tools (*see* Chapter 12 on boring tools).

The assembly diagram in Fig. 13.3 should help you understand how the whole boring head goes together. The sliding head fits in a dovetail on the body and is moved over its range using an Allen key to adjust the lead screw. The boring head itself needs to be

Fig. 13.2 The dial allows very fine adjustment of the cut.

mounted on an arbor to suit the milling machine. This can be any commercial blank-end arbor, with a suitable thread and register cut on the end.

The body of the unit can be made from 2in diameter EN1A, which leaves a small allowance for final machining. It should be threaded to be a close fit on the arbor. Take care to get the dovetail central and observe the depth

and width of the dovetails as closely as possible.

The gib strip (Fig. 13.6) should be left unfinished until the body and slide are completed, but it is useful to make the 'embryo' strip at an early stage, as it can be used as a gauge when fitting the two dovetails – the gib should just slide freely into the gap between the two parts, without any binding. In use, adjustment screws hold the gib, and therefore the two dovetailed parts, firmly in alignment.

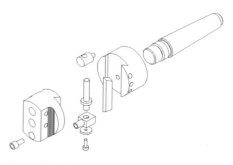

Fig. 13.3 Exploded view.

Fig. 13.1 The micrometer boring head.

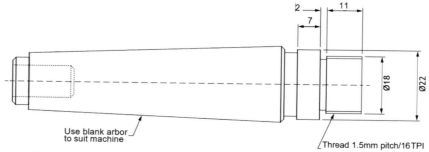

Fig. 13.4 Shank, based on a suitable blank arbor.

Use blank arbor to suit machine

2 7 11

Ø18 Ø22

Thread 1.5mm pitch/16 TPI

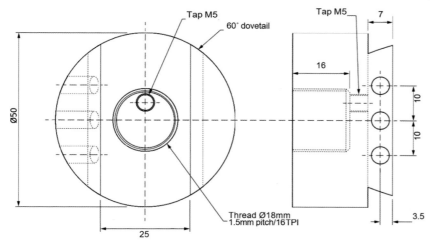

Fig. 13.5 Body.

The lead screw is a nice exercise in screwcutting but will benefit from finishing with a die. Note that M6 is 1mm pitch, but you can use other pitches if they suit your needs better. Do not omit the small slot – this will

be needed to lock the lead screw for final assembly.

The lead screw nut threads into the body. It is best to fit it and check its orientation before drilling and tapping it for the lead screw.

The slide (Fig. 13.9) is the most complex part of the boring head. Rough-machine the dovetail, then 'tweak' it so that the embryo slide aligns well with the body and use the gib as a gauge to finish the dovetail to width. Break the corners of the dovetails so they do not bind.

The two holes for the boring bars (Fig. 13.10) should be finished with flat-ended D-bits to get a close fit on your bars. Both 10mm and 3/8in sets are

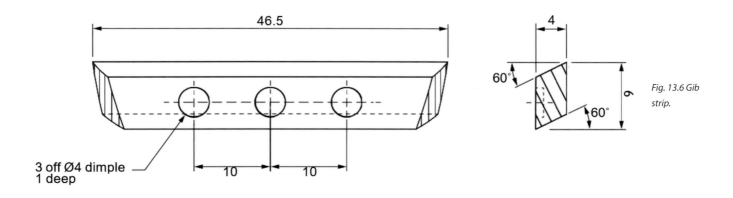

3 off Ø4 dimple
1 deep

Fig. 13.6 Gib strip.

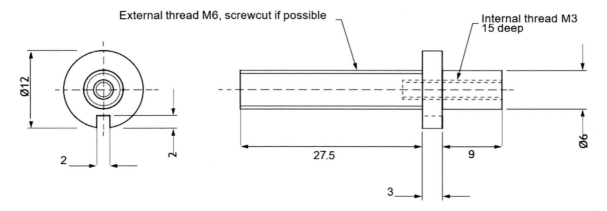

Fig. 13.7 Feed screw.

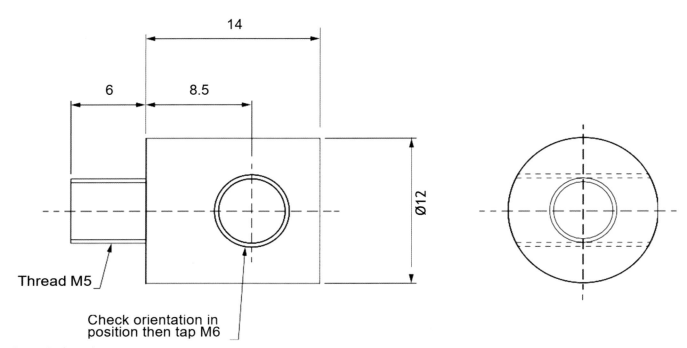

14

6   8.5

Ø12

Thread M5

Check orientation in
position then tap M6

Fig. 13.8 Feed screw locator.

available; it is probably wise to order and measure your set before finishing the holes to make sure they match your bars! The scalloped sides are not only decorative, but also reduce out-of-balance weight – the exact radius is not important.

The index dial, shown in Fig. 13.11, is a nice little job, and it is not too hard to make the 40 division marks. First angle the top slide to turn the taper and then use it to move a sharp pointed tool across the dial to cut the grooves. Dividing the 40 divisions can be done by fitting a 40-tooth change wheel to the back of the lathe spindle and using a springy strip of brass to improvise a detent.

The lead screw support should be carefully made so that it fits in the end of the lead screw groove on the underside of the slide, locating in the 7mm

hole. Make some careful measurements and ensure that the 6mm hole aligns exactly with the thread in the lead screw nut.

To assemble the head, you will need some M5 grub screws for the gib and bar clamp screws. Flat or pointed tips are both OK, but do not use those with serrated or cupped ends. Before fitting the lead screw, fully tighten the gib and take a light skim over the whole body and slide to bring them to their finished 50mm dimension.

Fit the lead screw nut and lead screw, then over these fit the slide. With the lead screw wound in a little, you should be able to fit the lead screw support (with an M5 cap screw), then wind the lead screw out. Fig. 13.13 may help with visualizing this. You may need to make small adjustments until the dial (when fitted with a 10mm long M3 cap screw)

gives a shake-free fit for the lead screw in the support. Use a screwdriver in the notch in the lead screw collar to stop it turning.

Check everything works, but bear in mind that you will need to use some retainer on the dial fixing screw on final assembly. A little moly grease on the lead screw and slides is a good idea.

The boring head should be used with the gib set fairly stiff, and the rpm rather slow (Fig. 13.14). The minimum hole size is about 12mm using the central hole and over 60mm using the outer hole.

The biggest challenge, as with any such head, is setting the boring bars just right, as it is very easy for them to rub. This is not an automatic head, so after each run through a hole, stop the machine and put on the cut. One

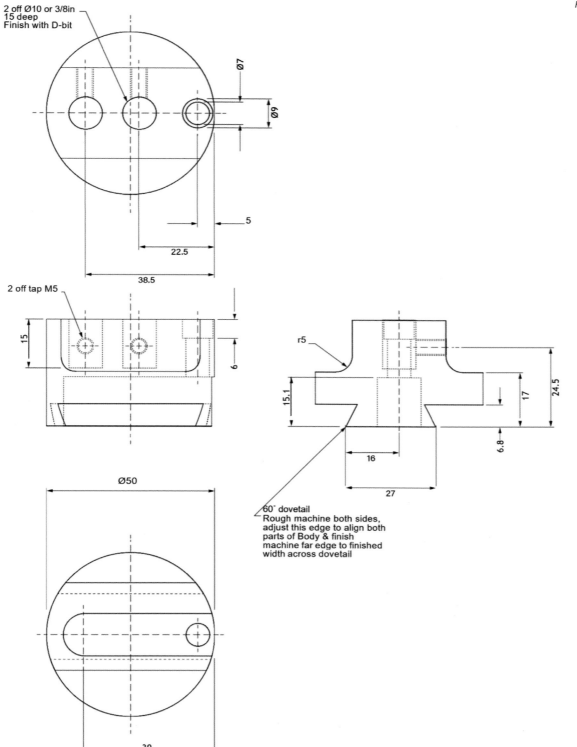

Fig. 13.9 Slide.

Fig. 13.10 The slide assembled to the body.

Fig. 13.12 Graduating the dial.

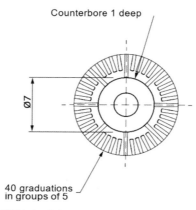

Counterbore 1 deep

Ø7

40 graduations
in groups of 5

Fig. 13.11 Index dial.

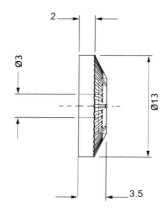

2

Ø3

Ø13

3.5

division is 0.025mm or 0.001in on radius, or twice that on diameter. With a little practice, you should have no trouble using the head to bore out holes to an accuracy of better than half that.

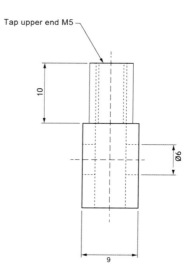

Ø7

6

14

13

Fig. 13.13 Feed screw nut.

Tap upper end M5

10

Ø6

9

Fig. 13.14 Rendered version in Alibre Atom 3D.

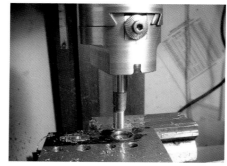
Fig. 13.15 The boring head in use.

14 A Box Tool

One feature of an engine I made was a cylinder that sits on top of four columns, each of mild-steel rod, turned down to a shoulder and threaded 7BA at the ends. While it was not particularly difficult to do this job, some care was needed to make the turned sections of consistent size and length. Such a task is much easier using a 'box tool' mounted in the tailstock (Fig. 14.1). This accessory incorporates a toolbit and a steady, so that it supports small work close to the cut, allowing cleaner, faster work to be carried out. Usually, a depth stop is fitted, enabling rapid, repeatable cuts. It is particularly useful when multiple parts must be made, such as a batch of custom screws from hexagon stock.

Bearing in mind that the full depth of cut must be taken in one go, with the tool held by the tailstock taper, this version is designed to complement smaller lathes. The design is rather smaller than usual, but with reasonable capacity (about 13mm). The limit is the depth of cut that can reasonably be applied, but this allows a rather simpler construction.

More complex versions use a toolholder located in a groove and held in place by three screws with toolbit adjustment by a fourth screw. This simplified version uses a single screw to both lock the holder in the groove and allow for adjustment (Fig. 14.2). The steady holder is simpler than usual as well, using a short section of 1/4in diameter material. This has the advantage that it is easy to make steadies in different materials (such as steel, brass or even nylon) to suit the job in hand. By mounting the steady on a swivelling base, it is simple to adjust for small errors in hand-cut notches in the end of the steady.

Few of the measurements are critical, but Fig. 14.3 is a guide. What matters is that the top surface of the toolbit is aligned with the centre line of the tool – if you achieve this, then you may vary the other measurements to suit what is in your scrap box.

MAKING THE BOX TOOL

The first job is a hollow MT2 taper arbor with a threaded nose (Fig. 14.4). This can be a shop-made one, but, although it is often less bother to buy a pre-turned arbor with a blank on the end, it may be easier to put a through hole in a shop-made one. I used the 60° UNF thread form, as my lathe was already set up for this, metric or Whitworth threads of similar pitch will be fine. The hole right through the arbor allows long workpieces to be shouldered down. Thread this hole M10 for a short way at the narrow end of the arbor. This will allow the fitting of an end stop, facilitating the turning of shoulders to a repeatable length.

The most involved component is the backplate. I used a slice of 50mm mild-steel bar, although a slice of continuous cast-iron bar would be fine. Start by facing one side and boring the central hole, then threading it to a firm fit on the

Fig. 14.1 The box tool.

Fig. 14.2 View showing the steady and toolbit.

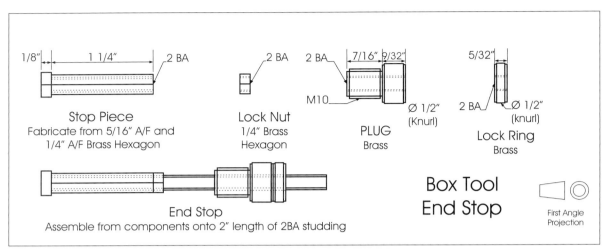

Fig. 14.3 Parts of the box tool.

Stop Piece
Fabricate from 5/16" A/F and
1/4" A/F Brass Hexagon

Lock Nut
1/4" Brass
Hexagon

PLUG
Brass

Lock Ring
Brass

End Stop
Assemble from components onto 2" length of 2BA studding

Box Tool
End Stop

First Angle
Projection

Fig. 14.4 Turning the taper shank. A blank arbor could be used.

Fig. 14.5 Using a four-jaw chuck to drill an offset hole.

Fig. 14.6 Tidying up the body when fitted to the shank.

arbor. Assemble the two parts, mount it in the headstock and face the front and clean up the circumference (Fig. 14.6). For my lathe, this required using a MT2-3 sleeve. Mark the back face of the backplate and arbor to show how they will align on final assembly.

Colour the front of the backplate using a magic marker or marking blue, and mark out the various holes and grooves. You can transfer it to a milling machine if you have one or mount the backplate on the vertical slide to use the lathe. Use a spacer behind the work to avoid the risk of marking the slide when a tool breaks through. Mill out the 5.5mm slot for the tool block mounting screw first

(Fig. 14.7), then a 1.5mm deep groove for the tool block. If you do not do this in order, you may have to remove the work, re-mark it, then fiddle to mount it to cut the slot parallel with the groove. The groove should be a good sliding fit for a piece of matching mild-steel bar.

Now make a centre pop at the pivot point for the steady and put the backplate in the four-jaw chuck. Centre the work on the pop, and drill for an M5 screw. Now bore a 13mm diameter recess for the steady block concentrically with the hole. Finally, assemble the backplate on to the arbor using Loctite, or a similar retainer, ensuring that the marks made earlier are aligned.

Fig. 14.7 Milling the tool block in the lathe.

The tool block and steady block should initially be squared up and trimmed to length in the four-jaw chuck. The M5 threaded hole in the steady block is best done with the work still in the chuck after turning the 1/2in register.

Marking for the various holes and slots is best achieved using a surface gauge set to height above a flat surface (Fig. 14.8). You can use either of the gauges described earlier and a piece of thick glass or even granite worktop will serve as a passable surface plate. Use the vertical slide to mill the tool slot. The accuracy of this slot is critical if the toolbit is to be properly aligned! I used a three-flute mill to make a slot the size of the toolbit, then took another 1/32in off the top of the slot. This allows for a little packing to raise the toolbit, should it be ground or wear a little low. M4 screws for the tool-clamp screws will be sufficient, but there is enough metal to use M5, if you wish.

The simple toolbit in Fig. 14.9 can be ground from square-section high-speed steel. Cutting off a short section of HSS bar is easier than it appears. First, cut a shallow groove all round, using a carborundum cut-off wheel in a mini drill. Grip the longer end of the bar in a vice. Cover it with a cloth to stop any splinters flying and give the other end a judicious tap with a hammer. Bingo! The angles for grinding the cutting edges of the tool are up to you, but I usually use the 10-degree for each angle, as shown in the drawing.

Fig. 14.8 Marking out the tool block.

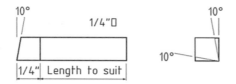

Example Toolbit HSS

Fig. 14.9 A suitable toolbit.

Fig. 14.10 HSS toolbit blank.

This is easy to judge for offhand grinding. In my limited experience, a good keen cutting edge, with adequate front clearance, is more important than having a different angle for every different material.

The hole for the steady should be accurate – ideally, it should be reamed with the work in the four-jaw chuck. Various materials have been suggested for steady rests. Conventional steadies usually use hard materials, such as phosphor bronze or hardened silver steel or gauge plate. These are particularly useful when shouldering threaded rods, which might otherwise cause heavy wear. As long as the working surface is polished, then the risk of scratching the work is minimized. Some recommend softer materials; brass could be suitable when turning steel, and even plastics such as nylon may be used. In truth,

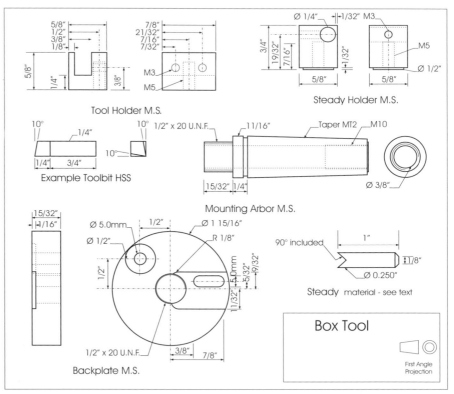

Fig. 14.11 End stop details.

Fig. 14.12 Parts of the end stop.

any material that is dissimilar from that being turned, is strong enough for the job, and is capable of being finished to the right shape and size will suit. You can experiment with a small collection of steadies of different materials. Note that the steady rods are tapered at one end so they can be reversed for small work.

The ability of the box tool to turn to a consistent diameter is complemented by a depth stop, which allows repeatable lengths to be turned. First, make a simple plug from brass rod, threaded M10 at one end, and knurled at the other and with a threaded hole through the middle. A section of studding runs in this hole, with a simple lock nut made from a knurled and threaded slice of brass.

The stop itself is made of two pieces of brass hexagon, drilled and threaded, and soft-soldered together using a 'keyring' blowtorch. The end piece could be from round brass, but the hex was a good fit! A lock nut was made from a slice of the smaller hexagon.

USING THE BOX TOOL

First, set the depth stop. The end stop is screwed on to the rod to give approximately the right length (Fig. 14.13), then the plug is screwed into the mandrel and the rod adjusted to the exact length. This approach ensures that the minimum of the studding protrudes from the end of the plug, maximizing the available tailstock travel.

Hold the material to be shouldered in a suitable chuck and mount the box tool in the tailstock. Run the barrel out a few turns to make space to reach the adjusting bolts and lock the barrel. Loosen the two blocks and the steady rod. Position the tailstock so the work goes past the steady, but not the toolbit, and lock it in position (see Fig. 14.14). It is now easy to align the steady rod into firm contact

Fig. 14.13 Assembled end stop.

Fig. 14.14 The box tool in use making a run of phosphor bronze bushes.

Fig. 14.15 Turning spigots on brass bar.

with the work. Adjusting the depth of cut for the toolbit is less easy, but can be facilitated by replacing the work with a rod of the same diameter as the desired reduced portion, and adjusting the tool to graze the surface of this gauge piece. Adjust the tool by moving the entire block, not by moving the tool in the block.

Cutting speeds should be similar to those for normal turning. Keep the steady lubricated, even if turning brass. Feed the box tool on to the work gently, until you feel it contact the depth stop. You may need to take deeper than normal cuts with this tool; it will take cuts of 11.5mm in phosphor bronze without any difficulty, and even deeper ones in brass. In such cases, however, you must be careful to feed the tool in very slowly, being very aware of the 'feel'.

15 A High-Torque Toolpost Spindle

It seems hard to believe that the idea of battery-powered tools was once impractical, as today they have become ubiquitous. The older NiCad-powered hand drills were particularly vulnerable to battery-pack failure, and it seems many workshops have at least one, if not more, of these in the scrap bin. They are also easy to find cheaply at car boot sales. Such a drill can provide the basis for a high-torque toolpost spindle for drilling and light milling.

The usual set-up for such a device is a two-stage concentric sun and planet or epicyclic gearbox, which reduces the motor speed from a few thousands of rpm to a hundred or so. There are several configurations that an epicyclic gearbox can take, but the standard type used for this purpose is a small 'driven' or input gear, meshing with and surrounded by three planet gears, which are mounted on a follower plate with three studs. In turn these three planet gears mesh with an outer 'fixed gear'. The reduction ratio from the driver to the follower is 1:1 + fixed gear/driven gear. By using small driven gears within a significantly larger fixed gear, ratios of around 5:1 are easy to achieve. Such gearboxes have several advantages: they are compact yet able to take a high-power input (because you have three times as many teeth in mesh) and the whole drive train is concentric.

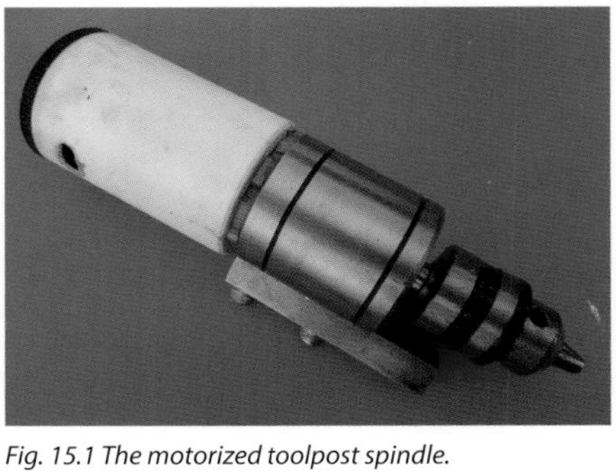

The cheapest and simplest 'screwdrivers' often run off about 3.6 volts provided by a three-cell power pack. They have a motor of a modest size and a simple nylon gearbox that incorporates a moulded-in fixed gear. Their output shaft runs directly in the end of the gearbox and they have no thrust bearings. As such, they are not suitable for heavy-duty purposes, although they are a useful source of parts and could be used to provide a powered feed for small lathes such as a Peatol or Unimat.

The bigger drill/drivers are built to deal with heavier tasks. I had the remains of two other single-speed drills, both with fixed-ratio gearboxes. One of these dated from the 1950s or 1960s and was designed to run off a car battery. An article in a long past issue of *Aeromodeller* demonstrated how to remove one stage of the gearbox of this drill and convert it to a glow motor starter. This raised the possibility of using the gearbox as the heart of a powerful toolpost spindle. The drill had a glass-filled nylon gearbox with an accurate bronze bushing, but after investigation the problem of dealing with the gears cast into the gearbox limited the possibilities for making use of this.

This left the power unit of the other single-speed drill/driver, a relatively early pistol-grip unit. Fitted with a Jacobs Multicraft chuck, it turned out to be better suited for modification. The output shaft of the gearbox ran in a cast-iron bush, it had a ball thrust bearing and, instead of being integral with the box, the fixed gear was an accurate metal casting. It also had a standard 540-type motor. Much maligned because of their high current requirements, these motors are near bullet-proof and capable of working at remarkable power levels. This was a well-made quality unit and I considered using it just to make a

Fig. 15.1 The motorized toolpost spindle.

Fig. 15.2 Planetary gearbox from a portable drill.

toolpost-mounting bracket for the motor/gearbox assembly, however, there was significant play and end-float in the main bearing. This was acceptable for a drill or a screwdriver, but not for any side thrust loads, such as those that would occur in light milling.

It took me a while to happen on the discovery that the lock screw inside the chuck had a left-hand thread. What had been an immoveable screw came free in an instant, and even the chuck itself took only the minimum of persuasion to unscrew, once I had carefully locked the armature. From this point, the conversion was easy. Two screws removed a plate that held on the clutch setting ring. This in turn concealed a stepped arrangement for tensioning a fairly powerful coil spring. Removal of the spring liberated sixteen 4mm bearings from eight holes in the gearbox. The slipping clutch worked by these sprung bearings engaging with eight protrusions on the front of the fixed gear. Under excessive torque the bearings are pushed into the holes and the fixed gear slips. For a toolpost spindle, there is no need to use or replicate the clutch, so this was not retained.

The next item to remove was a C-clip on the output shaft. This held in a simple thrust bearing consisting of two washers with balls between, inside a circular cavity. I took great care not to lose these bearings.

Removing three self-tapping screws from the motor mount revealed the gearbox proper. Along with several large shim washers, the casing contained three nylon stage one planet gears on a cast follower, and three

steel/iron gears on the machined and fabricated final follower/output shaft. This output shaft ran in a bush moulded into the gearbox; this was the source of considerable sideways play. Unlike the thrust bearing and gearbox, which were still well packed with grease, the bush was completely dry. Whether it had worn, or had been made oversize, was not obvious, but the fit was certainly more than just 'free-running'. The shaft itself measured exactly 9.98mm and appeared unworn (there was a significant section for comparison outside the bush). In a box of useful bits and pieces, I found two Oilite type sintered bronze bushes, nominally 10mm bore; these gave a much closer but still free-running fit.

It was clear that this would be suitable for a serviceable toolpost-mounted device, if the gearbox housing was rebuilt in steel and better-quality bearings were fitted (*see* Fig. 15.3). I started with a length of 2in diameter free-cutting bright mild steel. I used a half-centre to support the work for facing each end square, but had to do the rest of the machining with the piece supported solely by the chuck, so I avoided heavy cuts. After squaring off both ends I opened up the bore to 13mm in several stages, then I used a boring bar to open the bore to an easy fit for the bronze bushes. I then bored out a cavity for the fixed gear. The fixed gear is effectively free-floating in this space, centred by the other gears.

In the original drill, the gear acts as a clutch by engaging its eight 'protrusions' against eight ball bearings loaded by a variable spring. When the torque is sufficient to compress the spring, the

Fig. 15.3 New body casing with bearing bush and holes for gearbox.

gearbox slips. For this application, a slipping gearbox is unwanted, so instead of the bearing and spring arrangement I made the gear space a little longer than the gear (less its protrusions) and, using a rotary table, drilled eight holes around its periphery for the protrusions to permanently engage. While the unit was on the rotary table I drilled three motor mounting holes tapping size for suitable screws.

The next task was to bore out a space at the other end of the work for the thrust bearing. This was a cut-and-try task, requiring the repeated assembly of both bushes, the shaft and the thrust bearing. Both bushes had to be reduced in length, and the front one had to have its flange considerably reduced. They

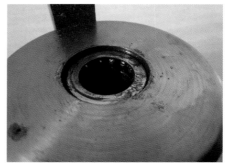

Fig. 15.4 Thrust bearing.

were of the porous Oilite type, so it was important to use a very sharp tool to avoid 'smearing' the cut surface and to prevent oil penetrating the bush.

The thrust race is a simple one: just two washers with either side of balls packed into the space between the shaft and surrounding cavity. I replaced the 3mm bearings in the thrust race with 1/8in ones – because I dropped the smaller ones on the floor and could not find them all!

Once I was happy with the overall fit, I epoxied the two bushes in position, using the shaft to ensure their alignment. I then shortened the overall length of the gearbox a little, to facilitate fitting of the spring clip that holds the shaft and bearings together.

Finishing off the gearbox was now largely a cosmetic job. I skimmed the outer diameter with a sharp tool and brought everything to a high polish with very fine wet and dry paper, followed by Scotchbrite pad. Then I put in two decorative grooves with a 1.5mm (1/16in) parting tool.

The two final tasks were performed in the milling machine. Protecting the ends with some alloy shims, I used a fly cutter to put a 16mm (5/8in) wide flat on one side, and drilled two holes, later tapped M6, on the flat. This would provide an attachment for a mounting bar, which I made from 13mm (1/2in) square stock.

After a trial assembly I saturated the output bushes in oil, then assembled the gearbox and thrust bearings, filling

Fig. 15.5 *Electrical connections.*

them roughly one-third with Finish Line Teflon grease, a product aimed at cyclists. It works really well and, being white, seems less messy than moly grease.

The motor has quite powerful permanent magnets and could easily attract a lot of swarf, so I made a cover from a curry powder pot. With a hole out of the bottom and three suitable holes it was tricky but not impossible to fit it using the motor mounting screws. I then cut it to length with a razor saw. Very simple electrics were a power socket, an on–off switch and a pair of LEDs, red for power on motor off, and green for run.

The tool can be mounted along the length of the lathe bed, or across it. In order to allow a reasonable amount of working space, I made the mounting bar longer than the gearbox casing. It is worth remembering that, in company with a swivelling top slide, holes can be drilled at almost any angle (*see* Fig. 15.6). Small slitting saws work very well with the device.

For test purposes I used a 12V (18V off load) 1.7A DC unregulated supply.

Fig. 15.6 *Vertical slide suitable for use with the spindle.*

This proved ample to power a 3/32in FC3 cutter through brass and then mild steel, cutting a full-width slot. Although this was not very demanding, the object of the exercise was to cut down on swaps between the lathe and the rotary table on the mill and simplify minor tasks such as putting a hexagon on small fittings. The motor is now better served by a more powerful, variable, pulse-width supply: a modulated 12V 4A.

Mounting the device on a vertical slide allows accurate centre-height setting, and also the milling of vertical surfaces or slots.

Despite the relatively modest speed, this arrangement has proven satisfactory for light milling tasks and has brought a new degree of convenience to work on the lathe.

16　Drilling and Reaming

Most people are familiar with the use of drills and drilling machines, but some advice on getting the best results might be useful, particularly for beginners who can find it very frustrating trying to get accurately sized and positioned holes.

As a first principle, unless you are drilling very small holes in large stable workpieces, always use a vice or clamp to keep the work in place under the drill. It is not just a question of keeping things safe; it is also essential practice to ensure maximum accuracy.

SELECTING EQUIPMENT

Generally, when choosing a drill, you will be off to a good start if you look for a decent-quality TiN (titanium nitride) coated set in 0.5mm increments. A full set from 1mm to 10mm in 0.1mm increments is a great asset, but if that is outside your budget, go for a smaller set and buy additional sizes when you need them for purposes such as tapping drills or pilot holes for reaming. The technique of sharpening drills is beyond the scope of this book, but it is well worth buying (or making) an accurate drill sharpener – quality drills are an investment that will last many years if looked after properly.

Basic drill presses typically come with a single chuck fitted directly to

Fig. 16.1 A selection of Jacobs-type chucks on arbors.

Fig. 16.2 Keyless chuck fitted to a lathe.

a tapered plug on the spindle. Larger ones, milling machines and lathes have an internal taper and can take interchangeable chucks fitted to a suitable arbor. The most affordable chucks are keyed ones, like those in Fig. 16.1. This selection of chucks covers several different size ranges. If you only have one chuck, a 1mm to 13mm is probably the most useful size, as this will hold both small (but not very tiny) drills up to the capacity of most benchtop machines.

It is well worth getting a decent-quality keyless chuck, such as the one in Fig. 16.2. They have the advantage of tending to self-tighten, so they are less likely to let the drill slip under a heavy load and are also a genuine pleasure to use. It is probably best to avoid very cheap plastic-bodied keyless chucks – like those that are often found on pistol drills – as they rarely offer the accuracy and grip of a proper full-size keyless chuck. If a keyless chuck tightens itself

Fig. 16.3 Tommy bar holes on keyless chuck.

in use, you should be able to find a hole that allows you to loosen it using a C-spanner or a snug-fitting bar as a lever.

Whatever type of chuck you get, it will probably need to be fitted on to a taper arbor. This arbor should match your machine, so typically a Morse taper MT2 is most usual. The taper at the chuck end might typically be one of the B-series (for example, B12) or a

Fig. 16.4 Arbor showing taper for fitting chuck.

Jacob's taper (for example, JT6) and your supplier should be able to supply a matching arbor (*see* Fig. 16.4). To fit the arbor, make sure the socket in the chuck and the matching taper are completely clean. Wind the jaws of the chuck right in and rest the 'nose' of the chuck on a wooden block. Insert the arbor into the chuck and tap the end of the arbor with a wooden mallet. Do not wallop it. If you do not have a wooden mallet, find another piece of wood to protect the arbor and use a medium hammer. If you ever need to part the chuck from its arbor, you can get pairs of matching wedges that are squeezed in from the sides.

To use the chuck, simply push it into the (clean) socket with firm hand pressure. Under normal loads the chuck should not be in any danger of slipping, as the load is axial and pushes it into the socket. On most drilling machines you can eject the chuck using a tapered wedge through a slot in the barrel.

DRILLING A HOLE

Drilling a hole is as simple as starting the machine and running it at the right speed. Most people give scant regard to drilling speeds, and a few hundred rpm will do for most items, but you will be more productive if you change speeds. To create the hole, the drill is fed in

using a feed lever or capstan. If the drill starts to 'squeal' or you feel increasing resistance, back the drill off and clear any swarf from the flutes.

The feed rate for drilling holes is best found by feel. It is often hard to run the work fast enough for small drills, in which case you will probably find that you need to take your time, to avoid damaging the drill.

Although it is true that drilling a hole is as simple as fitting the drill and feeding it into the work – and doing this will usually give you a hole – you will want to maximize the chance of the hole being accurately on centre and to size. To do this it is best to start off using a rigid spot drill or a centre drill (Fig. 16.5). Spot drills are generally better for this task as they are less fragile in smaller sizes. You just need to make enough of an indentation to take the tip end of your normal drill, so it only has to cut on its flutes.

For really large holes, it is worth having a few equally large drills. 'Blacksmith's drills' have a reduced shank to fit in a normal chuck, but I also have a few large drills with Morse taper shanks (Fig. 16.6). It is rare that you will want a hole the exact size that such a drill gives, but they offer a great way of removing

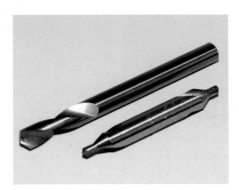
Fig. 16.5 Spotting drill and centre drill compared.

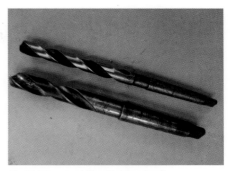
Fig. 16.6 Oversize drills with taper shanks.

metal quickly in preparation for boring an accurate hole.

REAMING

No matter how good your drills are, you should not expect them to give you perfectly accurately sized holes. Drills are designed to prioritize rapid removal of material over exact size; in fact, they will usually drill slightly oversize. In contrast, reamers (Fig. 16.7) are designed to remove small amounts of material but to give well-finished and accurately sized holes. They are used to open out undersize holes to finished diameter. Reamers are available in different tolerances. The commonly supplied H8 tolerance will give a close sliding fit on stock material and is a good general-purpose option.

Hand reamers have a gentle taper on the end and typically have a square to allow them to be turned using a (sizeable) tap wrench. Machine reamers do not have a taper and usually have a morse taper shank. In practice, with care it is possible to use both types of reamer in a lathe, although a hand reamer will need to be held with a drill chuck.

Reamers need to make a proper cut to work well and avoid excessive wear,

Fig. 16.7 Machine reamer.

Fig. 16.8 Small reamer in use on brass.

so the hole to be remade should not be too close to finished size. A pilot hole for reaming should ideally be about 0.1 to 0.4mm undersize, the smaller figure for holes of a few millimetres increasing to the larger figure for 25–30mm.

Cutting speeds for reamers are about a third of the speed you would use for a drill of the same size. Although they only remove small amounts of material, small reamers can clog easily so they should be withdrawn regularly during a cut. Never, ever, reverse the machine when withdrawing a reamer as this will rapidly blunt the tool by 'rounding over' the cutting edges.

Tapered reamers are used for making precision holes, such as sockets for Morse taper arbors. They must be used only for finishing holes.

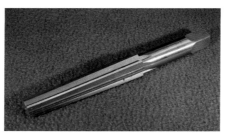

Fig. 16.9 Taper reamer for finishing or repairing Morse taper sockets.

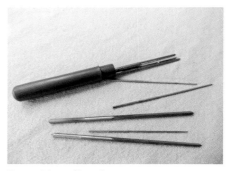

Fig. 16.10 A set of broaches.

BROACHES

Hole-making broaches resemble reamers but are always tapered. Instead of flutes they typically have three flat faces, with the cutting performed by the edges where these faces meet.

17   Taps and Dies

TAPS

Tapping is the process of cutting or forming an internal thread using a tap. This is a form tool in the shape of a hardened rod screwed on the end with a number of flutes, creating a series of cutting edges.

Most quality taps come in sets of two (taper and plug) or three (taper, second and plug). Taps typically all produce a full thread form but have decreasing amounts of taper or lead on the end. For a deep hole, it is wise to cycle through the taper, second and plug taps in turn, rather than trying to do the entire hole in one go. Some taper taps may cut a thread that is considerably undersize, and these cannot be used to finish a through hole.

Normally, the end of the shank has a square for securely holding the tap when in use. It is also possible to get fluteless taps that work exclusively by extrusion. These are particularly useful for materials such as stainless steels, which can be difficult to thread using normal taps because of their tendency to work harden.

To create a thread using a tap you need to start with a suitable hole. Typically, a hole size that gives a thread depth of around 60 per cent of the theoretical full thread form is used – that is, rather larger than the 'core diameter' of the tap. Smaller holes greatly increase the risk of tap breakage, while making little practical difference to the strength of the thread. Tables of tapping sizes are widely available, but beware of those that recommend very high thread depths.

The tap needs to be held in a suitable tap wrench. The commonest style is a T-handle with various clamping arrangements for gripping the square section on the end of the shank

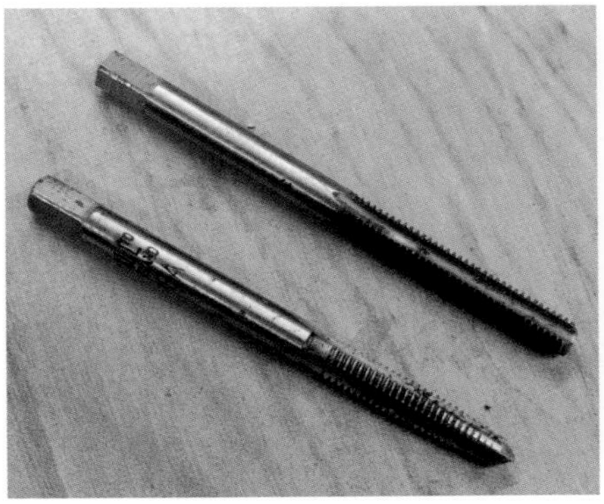

Fig. 17.3 Tap fitted to tap wrench.

(Fig. 17.3). Some small tap wrenches have collets and hold the tap by its round shank; this can be an advantage when using small taps as they can spin in the collet if overloaded. For the very tiniest taps it can be helpful to use a pin chuck. Tap wrenches can also be used for turning hand reamers and similar tools.

Gradually screw in a series of taps to produce an increasing full thread form inside the hole. It is important to keep the tap aligned with the hole. A tapping guide is a simple jig, comprising a close-fitting hole in a square-ended block, to help minimize breakage of small taps and also ensure that threads are truly perpendicular to the surface.

The ideal technique is to advance slowly, making full turns of the tap, then turning backwards to break off the swarf that forms into short pieces. This process can be eased by using a lubricant such as cutting oil or

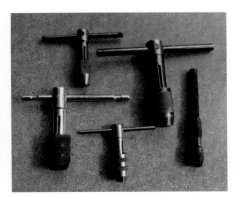

Fig. 17.1 A taper and plug tap.

Fig. 17.2 A selection of tap wrenches.

specialist tapping fluid (a cutting lubricant formulated especially for thread-cutting operations. Most tapping fluids also serve a useful purpose for drilling and reaming.)

Do not try to tap too deeply in one go, as the flutes will rapidly fill with swarf, and be prepared to clean out the hole if necessary. For deep holes, it can be beneficial to run the taper tap in until it starts to bind, then follow with the other two taps in turn. You may then go back to the taper tap and deepen the thread further.

Serial taps similarly have increasing amounts of taper and have different diameters. The smallest size is a 'rougher' and with low thread depths may do little cutting beyond establishing the path of the thread in the hole. These are marked with a single ring on the shank. The next, intermediate, tap is marked by two rings. The 'finishing' tap has no rings, and it cuts the full thread form. The increasing diameter of serial taps generally require less cutting force; they can also last longer. If you use only the rougher and intermediate taps, a quite tight thread is created, which is ideal for threads for adjusting screws.

Spiral taps are made with spiral flutes and have better swarf-clearing properties for use under power.

Taps are available in almost as wide a range of thread forms and dimensions as there are threads, however, beyond about 25mm diameter, using taps and dies becomes increasingly impractical. Because of this, large threads are generally made using machine tools.

DIES

Dies are used to cut external threads. They essentially work in the same way as taps but, because they work externally rather than inside a hole, they are less fragile and less vulnerable to swarf build-up. This means only one die is needed for any thread size.

Adjustable dies have a slot allowing them to be slightly closed or opened up in a diestock to achieve accurate sizing of the thread (Fig. 17.6). The teeth of a die are relieved so that the first few teeth do not cut to full depth. Sometimes, dies are relieved on both sides, but normally the side stamped with

Fig. 17.6 A traditional split die.

the size of the die is fully relieved, while the other is only partially so. This can be used to advantage when a thread right up to a shoulder is needed. In this case, the thread is cut normally, then the die is reversed and run up to finish to the shoulder.

Unsplit dies that cannot be adjusted in size have become more common, as industry wants solutions that do not need time-consuming adjustment. As they are required to be very accurate, they are often more expensive than adjustable dies and often marked to show the degree of fit they produce (Fig. 17.7). Beware cheap hexagonal die nuts; these are generally intended for truing damaged threads, turned by a spanner and often struggle to cut a new thread accurately.

Fig. 17.4 Serial taps, used in order left to right; note the rings on the shanks.

Fig. 17.5 A set of dies.

Fig. 17.7 Unsplit dies, not to be confused with cheap die nuts.

A diestock is used as a holder for thread-forming dies (Fig. 17.8). A diestock has a cavity slightly larger than the nominal size of the dies, and a series of screws around its periphery. One of these is used to centralize the die, while on the other side is a cone-tipped screw used in the gap of the die to open it out. Adjacent to this screw are two more, which engage with dimples or slots in the die to close it up. A diestock for unsplit dies may only have a single screw. Because of the need for adjustment, threads are best started with a split die slightly opened, then closed up after a trial until it cuts to size.

A tailstock die holder is a device for supporting thread-forming dies from the tailstock of a lathe. Although not guaranteeing as accurate a thread as screwcutting, the use of a tailstock die holder is superior to cutting threads with a standard hand-held diestock.

Fig. 17.8 Die in a diestock.

Fig. 17.9 Tailstock threading attachment for a lathe.

It typically has the same adjustment screws as a diestock and an arrangement to allow the body to slide back and forth. In the absence of a tailstock die holder, it is sometimes possible to support a standard diestock from the tailstock.

As with taps, use a cutting or tapping fluid for best results. Working the die by rotating it a full turn forwards, then half back, is still good practice, but, as swarf can fall free of a die more easily, it is not essential.

If you have an unidentified thread and you want to buy fixings or cut a thread to match, you need to know its pitch and thread angle. This can be difficult to estimate accurately by eye. Aside from experimenting using fixings with known threads, you can use a screw pitch gauge to find a match. These are sets of metal blades profiled to match each of a series of screw threads. The gauges can be used to test the pitch and thread form of both screws and larger internal threads.

Imperial threads usually have their sizes given in threads per inch (TPI) and their nominal outside diameter in inches. Metric threads are specified by outside diameter (M6, M5, and so on) and the pitch between thread crests is given in millimetres.

DIE (SCREW) PLATES

A die (screw) plate provides another method of forming small external threads. It is a flat, hardened metal plate with numerous screwed holes, and is used for forming small threads on relatively soft metals (Fig. 17.10). Screw plates are commonly used in watch-

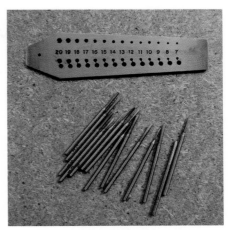

Fig. 17.10 A screw plate and accompanying taps for sizes M2 and below.

making, where the dimensions are too small to make the use of conventional dies practical.

If you have suitably sized dies, it is possible to make your own screw plates from gauge plate. Tap several holes near the edge of a relatively thin piece of gauge plate, using a fine saw to make a cut breaking into each hole. Clean up the holes by running a tap through again, then harden and temper.

MAKING TAPPING GUIDES

Tapping guides are relatively simple to make and will rapidly repay the effort invested in terms of more accurate holes and, with small taps, reduced breakage.

If you want a multi-purpose tool, the guide shown in Fig. 17.11 is ideal. Made from a length of mild steel or aluminium bar, or even 3D-printed, like this example, it has a series of holes drilled to match the shank diameters for a series of taps. Adding a taper along its length can make it more convenient to

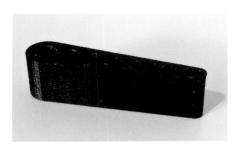

Fig. 17.11 3D-printed tapping guide for a range of small sizes.

use. As the guide does not need to be a close fit, a series of holes in increments of 0.5mm will provide for both imperial and metric sizes.

The ideal solution is a set of 'top hat' guides, like the example in Fig. 17.12. Each one is drilled clearance for a specific tap. The expanded base makes it easy to hold in position on larger surfaces, but it can be reversed when tapping a hole in a cramped location. A set of these that match a box of taps can be turned up from offcuts in half an hour. They could even be 3D-printed as they are not subject to significant wear.

Fig. 17.12 'Top hat' tapping guide in use.

18 Multi-Tools

Multi-tools live in an interesting area between machine tools and hand tools. These tools basically comprise a hand-held powered spindle, usually with variable speed, and fitted with a collet. The 'multi' in the name reflects the bewildering array of different fittings that can be used with them.

Typified by the Dremel brand, multi-tools are available in a wide range of styles and sizes, from slender low-voltage versions to hefty mains-powered types. A new generation of high-power lithium battery-powered multi-tools has recently come to market; these are barely larger than the traditional smaller-sized low-voltage units, which means they can be used in otherwise inaccessible places. Innovations include integral work lighting. Some of the larger mains-operated versions now come with a flexible drive to a slender hand-held unit, allowing a high level of power to be combined with an ability to be used for delicate tasks.

Some multi-tools, especially cheaper ones, are prone to overheating if used continuously for a long time. For this reason, it is good idea to rest the tool for several minutes whenever its starts to feel warm. Avoid using excessive pressure, causing the tool to slow down significantly, as this will reduce the flow of cooling air and increase current consumption and heating.

Fig. 18.1 A lithium battery-powered multi-tool with its charging station.

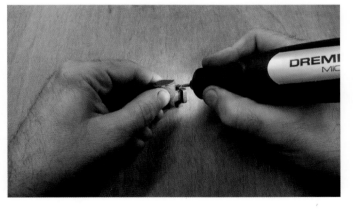

Fig. 18.2 Multi-tool with integral work light.

When you buy a multi-tool it will probably come with a selection of different tools, usually with 1/8in shanks. These tools are suited to a number of different tasks.

METAL CUTTING

Various types of cut-off wheels are available, but they are usually disposable and mounted on to a simple mandrel

Fig. 18.3 Fibre-reinforced cutting disk.

Fig. 18.4 Cutting disk and arbor.

DRILLING

Tiny carbide drills with 3mm shanks are often included in sets of accessories. These are excellent for drilling printed circuit boards, but must be used with some form of drill-press accessory – they are far too fragile to be used hand-held. Close inspection of such a selection will often reveal what is actually a random grouping of small engraving bits and milling cutters as well as micro drills.

Some drills are supplied or can be fitted with drill chucks allowing them to hold a wider range of small drills. In general, you will have more success using these small chucks (they screw on to the end of the spindle in place of the collet closer) to hold small-diameter HSS drills. Down to diameters around 1mm, it is possible, with care, to use HSS drills hand-held.

It is important to realize that these small chucks are very liable to loosening off rapidly when exposed to vibration or sideways loads. This means they are really only suitable for use with drills

(Fig. 18.3). The very thin, brown carbide wheels are excellent for cutting small, hard wires and even small pieces of high-speed steel. However, they are rather fragile and it is not unusual to get through several on a larger cutting task. That said, the ability to cut hard materials in inaccessible places is probably the greatest benefit a multi-tool brings to any workshop. More robust fibreglass reinforced wheels (Fig. 18.4) are preferred for jobs such as cutting through seized-on bearing races or even rough-shaping and cutting tools of high-speed steel.

Fig. 18.5 These tiny drills and cutters are very fragile.

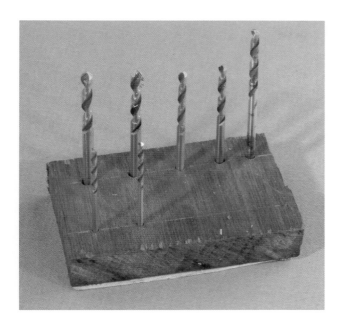

Fig. 18.6 A set of spotting drills.

Fig. 18.7 Various polishing mops and compound.

CLEANING AND POLISHING

Most accessory sets contain a variety of felt polishing mops and points, which often fit on a screw-ended mandrel. Felts are best used with polishing compound; this is often supplied in small containers of various grades, but you can also get excellent results on many metals and hard plastics by using automotive chrome polish.

Brushes can be made of steel, brass, or various types of plastic, and you need to choose the right one to suit the material you are working. For example, beware of using steel brushes on aluminium alloys as they can embed small steel particles that later rust and spoil the job.

Protection for the eyes and hands is essential when using these accessories, as they can eject sharp wires at great speed.

GRINDSTONES

Although many accessory sets include an assortment of different-coloured where the loads are all axial. Use with other types of accessories will almost always see the tool coming loose and potentially being thrown across the workshop.

Interestingly, what appear to be plain twist drills in multi-tool selection boxes often have a relatively fine 60-degree tip and a relatively short shank (Fig. 18.6). These are actually spotting drills and you will find them much better than centre drills for marking the locations of holes to be opened out with larger drill bits.

Fig. 18.8 A selection of rotary wire brushes.

Fig. 18.9 Nylon brushes are less abrasive.

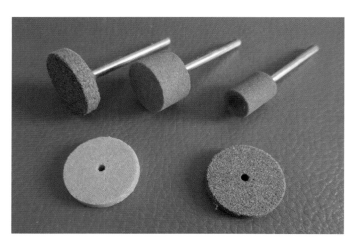

Fig. 18.10 Different types and grades of grinding stones.

Ideally, a set will indicate what materials each colour of stone is best for, but, if you have to experiment, unsuitable stones will reveal themselves by rapidly losing their shape. Again, eye protection is essential due to the risk of breakage.

SANDING DRUMS

Aluminium oxide sanding drums are intended for use with wood, but they are actually really effective on steel, cast iron and aluminium alloys, both for cleaning up castings and for blending corners into neat curves. If you run the tool nice and quickly and avoid forcing it, the drums are surprisingly long-lived and often more effective than grindstones, as they have a coarser grit size and cut faster.

BURRS AND RASPS

Most tools come with a selection of burrs and rasps, and often some diamond-studded points (Fig. 18.11). The coarse metal rasps are best used on wood and soft plastics, but this must be done with care, as they can remove material at a startling rate! In contrast, diamond-studded points can be used for delicate work on hard materials, even engraving glass. The point in Fig. 18.12 is fitted to a small specialist engraving tool that has a motor in a flexible mount. This makes it easier to apply gentle and consistent pressure when engraving. One particularly useful task for such burrs is cleaning up fine details on iron castings.

MULTI-TOOL SAFETY

As they are hand-held, it is easy to break tools in these machines – cut-off wheels and drills in particular. Another hazard is brushes throwing wires in unexpected directions and even small grinding stones shattering if they are damaged. All tools also cause swarf or small particles to be ejected. It is therefore essential to wear good-quality eye protection when using any multi-tool. Use a face mask when performing any task that will produce dust, and ensure that you have good ventilation in the workshop.

Some multi-tools can run very fast – always check that you do not exceed the maximum safe speed for any tool.

---

grindstones, these are often not as useful as they look. Generally speaking, they are not up to larger sharpening tasks, which are better carried out by taking the work to a bench grinder. However, they can be useful for activities such as removing moulding marks from castings.

Fig. 18.11 Steel and diamond-coated burrs.

Fig. 18.12 A diamond-coated burr in an engraving tool

Fig. 18.13 Fettling castings using a selection of burrs.

19   Basic Gear-Cutting Tools

Although the information needed to successfully produce your own gears is widely available, the processes tend to be buried in technical words and formulas. In this world where CNC (computer numerical control) is so prevalent, many hobbyists have become obsessed with accuracy and lack the confidence to produce satisfactory gears using what they see as modest equipment and skills. While there is some truth in the first of these issues, the maths and technical details can be made simple. On the equipment and skill front, making your own gears is both rewarding and surprisingly straightforward. As long as you take the process one step at a time, excellent results can be achieved.

Many people think it is necessary to have an expensive dividing head with a set of dividing plates to make gears. In fact, you can use a simple rotary table or a dividing head without plates (Fig. 19.2), as the accuracy of its setting

Fig. 19.2 Cutting a gear on a milling machine.

will depend almost entirely on the quality of the gears within it – any error in setting the dial is divided by the gear ratio. You may even use a simple lockable spindle with an improvised dividing device and an existing gear with the same number of teeth.

A similar argument applies to the generation of cutter profiles using circular form tools. Although several different approaches have been taken to calculating the dimensions to use, this 'simplified' approach can still produce gears with a profile that is within the tolerances demanded by industry.

For purposes such as timing gears, driving mechanisms and gearboxes, home-made gears should be more than adequate. If you model stationary steam engines, you may have been shocked by the prices of ready-cut bevel gears. However, with a little patience you should be able to make your own in no more than a few hours.

There are two basic requirements for making gears: a method for indexing the gear around one tooth at a time, and a suitable cutter to generate the tooth shape. There are more complex options such as hobbing machines that simultaneously rotate the gear blank as the cutter shapes the teeth, and dividing heads and rotary tables, but they go beyond the scope of the modest tools covered here. Fortunately, there are simpler ways of making the form tools for gear cutting.

The most consistent approach for making custom cutters is that of using hardened silver-steel 'buttons' to shape the sides of a cutter blank (Fig. 19.4). This process is somewhat involved and has been well described elsewhere, notably by Ivan Law. It produces cutters suitable for the 'one tooth at a time' approach that can be well within the tolerances required of commercial gears. It is an

Fig. 19.1 A gear machined in cast iron.

Fig. 19.3 A spiral hob for use on a gear hobbing machine.

Fig. 19.4 Using the button method to turn a gear cutter blank.

Fig. 19.5 Gears made using single-point cutters made by the button method.

approach that may be suitable when making quantities of gears, such as those in Fig. 19.5.

One method that appears to have had very little coverage is the simple expedient of hand-filing cutters to fit existing gears. It is surprisingly straightforward. The secret lies in regularly offering the embryo cutter up to the gear and holding it against the light to check the fit. The quality of the resulting cutter is almost entirely down to the patience of the person wielding the file. For obvious reasons, this approach is best limited to single-point cutters that can be made from small pieces of gauge plate. Pinions with small numbers of teeth sometimes need cutters with adjusted profiles, to prevent undercutting.

You do not need a pattern gear that is exactly the same size as the one you want to cut; it is sufficient for it to be in the same 'cutter range'. I have used this technique to make cutters for gears, including 63-tooth and 9-tooth gears using a 65-tooth gear and a 10-tooth pinion as patterns. Both gears were entirely satisfactory in use.

Taking the example of the 63-tooth gear (Fig. 19.6), I needed a number 2, 20-degree pressure angle, 1-module cutter. As I was only planning to make one gear in a fairly soft material, I decided that a complex cutter was not needed. I filed my cutter from a 1/2in by 3/4in piece of 1/8in gauge plate, offering it up to the 65-tooth change wheel as a template. I started by angling the end to provide relief, then roughed out a wedge shape with a smooth file. A half-round needle file made hollowing the sides of the cutter easy. Gauge plate is quite tough and files slowly, so it is easy to approach an accurate shape gradually. Once I had a good fit, I drilled

Fig. 19.6 Aluminium gear made using a hand-filed cutter profiled to match the adjacent nylon gear.

Fig. 19.7 Hand-filed gear cutter after hardening.

Fig. 19.8 Cutter mounted on arbor and honed on its face.

Fig. 19.9 End view of arbor.

the plate so that it could be fitted to a holder. The cutter holder in Fig. 19.8 is an MT2 arbor with two M6 holes in it, one for the fixing screw and one for a second screw as a stop (Fig. 19.9).

I hardened the gauge plate by heating to red hot then dropping it in sunflower oil. I then tempered it at about 170°C.

To use this tool to cut a gear you need some form of indexing device to hold the blank and index it one tooth at a time. If you do not have a suitable dividing device, you will need to make a simple spindle of some kind to support the gear and allow it to be indexed and clamped for cutting each tooth. Various designs for both simple and complex indexing and dividing heads have been published over the years. If you fit a wooden disk with a suitably marked paper scale on the other end, you can then use a simple pointer and index by hand. Somewhat better is using a gear (such as a lathe change gear) with a multiple of the number of divisions needed at one end and some form of simple spring detent.

For metric gears you only need to calculate three numbers, from the module (M), which determines the tooth size and is the distance between teeth divided by π (about 3.142) and the number of teeth. The results are in millimetres:

Pitch circle diameter (PCD) = N × M
Blank diameter = (N+2) × M
Depth of cut = 2.25 × M

For imperial gears the calculations are based on the diametric pitch (DP), which is calculated as π divided by tooth pitch. This is the inverse of module, and the number of teeth (N). The results are in inches:

Pitch circle diameter (PCD) = N/DP
Blank diameter = (N + 2)/DP
Depth of cut = 2.25/DP

The depth of cut may need to be adjusted to allow for any errors in your cutter but it should be fine if you have made it a good fit for a similar gear.

Pitch circle diameter is needed for setting out gears; for optimal meshing, two gears should be spaced apart by using the following formula:

Gear spacing = (PCD Gear 1 + PCD Gear 2)/2

To use the tool, transfer the blank, still in the three-jaw chuck where it was cut on the lathe, to a rotary table on your milling machine. The important points for cutting any gear are getting the cutter dead on the centre line of the blank and calculating the correct cutting depth for the cutter. Cuts are made with the cutter running at about 100rpm and advancing it very slowly across the blank, using brushed-on cutting fluid.

For softer materials, such as aluminium alloys, there may be some extrusion of metal at the crest of the teeth. This can be removed with a skimming cut off the crown of the teeth in the lathe, or by tidying up the edges of each tooth with a needle file.

I have used this approach successfully for other one-off gears, including achieving the 'worn' profile of the prototype gear in a model crane and its pinion (Figs 19.10 and 19.11), and even producing very tiny bevel gears (Fig. 19.12) for a steam engine governor.

Fig. 19.10 Gear and pinion cut using hand-filed gear cutters.

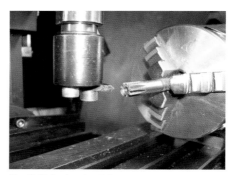

Fig. 19.11 Cutting the pinion.

Fig. 19.12 Tiny bevel gear cut using a hand-filed cutter.

20 A Quick-Change Toolpost

This quick-change toolpost will be suitable for almost any lathe of about 80mm (31/2in) centre height, including Mini-Lathes and Myford ML7 and Super 7 lathes.

Whenever you tackle a complex turning task, which requires you to repeat a series of operations on several parts, you may get frustrated by the limitations of a standard four-position toolholder. One problem is that, in practice, they normally only hold two

Fig. 20.1 The quick-change toolpost.

Fig. 20.2 A traditional four-way toolpost.

tools conveniently. One solution is a quick-change toolpost; a variety of suitable 'QCTP's can be bought for mini lathes, but once you factor in the need for a good set of compatible toolholders (I have twelve, and could usefully have twice as many), the costs start to spiral. An alternative is to make one yourself. Although this is not a raw beginner's project, it is a good test of your skills once you are more familiar with the lathe. Note that, although it can be made entirely using the lathe itself, milling the dovetails and toolholder slots is much easier if you have access to a milling machine. The project is also much easier if you have a metal-cutting power saw of some kind. You will need to cut significant quantities of quite large-section mild steel, and if you have to do it by hand, you should be prepared for some hard work.

MAKING THE TOOLPOST

Design

The design is based on that of the Nakamura Toolpost, but this has a couple of limitations: the piston does not retract automatically and has to be pushed back; it only takes one toolholder; and it overhangs the top slide excessively. The changes here aim to address the last two of these issues by adding a second dovetail and piston and reducing the overall dimensions of the block. Internal springs cause the pistons to be self-retracting. The toolholder offers ease of use and rigidity, the handle requiring very little pressure to fully lock the holders. The dimensions given allow the sliding and rotating parts to move freely without excessive slop and will help make sure

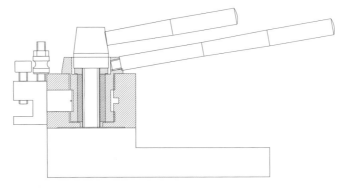

Attached to top slide using original clamp nut

Fig. 20.3 General arrangement of the toolpost.

Fig. 20.4 The unused piston is visible here.

the dovetail fits are right. In practice, use your experience and judgement bearing in mind the function of each pair of mating surfaces.

The operating principle of the toolpost may not be immediately obvious from the general arrangement. The main body (block) of the toolpost has a circular cavity in which a cylindrical cam is fitted. The cam has a hole through the middle, allowing a free central pillar to be threaded on to the toolpost stud and clamped by the original lathe clamping nut. The pillar is sized to clamp only the block, so the cam, with its own handle, is free to rotate. The eccentric portion of the cam bears on two brass 'pistons', one in each of the dovetails on the block (Fig. 20.4). Toolholders fit on these dovetails, and when the handle is used to turn the cam, it pushes on the pistons, locking the toolholders in place. Each toolholder has a setting button on top, which allows it to be removed and replaced at exactly the same height. To make using the toolpost easier, wire springs in the pistons engage with a groove in the toolpost block, so that when the cam is turned back, the pistons release automatically.

Materials

The following materials are needed for the toolpost itself:

◆ just over 11/4in length of 2in square bright drawn mild steel (BDMS) or continuous cast iron for toolpost block;
◆ about 11/2in of 11/2in diameter round BDMS for the cam;
◆ 2in of 5/8in diameter brass for the pistons;
◆ 3in of springy wire about 1mm/0.04in diameter for piston springs;
◆ 6in of 3/8in diameter BDMS for the handle;
◆ 11/4in of 3/4in diameter BDMS for the pillar.

For each standard or boring bar toolholder you will need the following:

◆ 2in of 1in square BDMS;
◆ 11/2in of M5 studding;
◆ 4 M5 cap-head screws;
◆ a short length of 1/2in diameter brass;
◆ 1 M5 nut.

Note that, in practice, you should include a cutting allowance. This means that, for example, 20in of bar will make nine full-length toolholders, and leave a useful offcut. Other types of toolholder may need different materials.

A final thought: toolposts for small lathes are often made from high-strength aluminium alloys. You may wish to use such a material for the block and toolholders, especially if you have a way to get them hard anodized.

Construction

It is not necessary to describe every detail of the construction, but the nature of some of the parts is such that it is useful to give a recommended approach that minimizes the number of set-ups required. The following process is not exactly how I made my holder, but it is how I would make a second.

The Block

The toolpost block is made from a single piece of 2in square mild- or medium-carbon steel. The former is easier to machine, but the latter should last a lifetime. Mount it in a four-jaw chuck, getting it a central as possible, face off the bottom to a good finish and turn the shallow rebate. The rebate is essential to ensure the outer edges of the block contact the top slide, maximizing grip. Now reverse the block in the chuck and face it to exact length. Drill the central hole in two or three stages. Use this hole to start your preferred boring tool to make the central 1in diameter recess, noting that it should be parallel-sided with a well-finished flat base. Now with a suitable recessing tool, put in the internal 1/8in groove. The centre line of the piston holes will align with this groove, so take care to get it in the right place.

Remove the block and mark out two faces with centres for the two piston bores. Return the block to the four-jaw chuck to bore and counterbore each piston hole in turn. You can drill these holes, but if you are not sure of the finish your 1/2in drill leaves, it might be preferable to use a reamer or finish

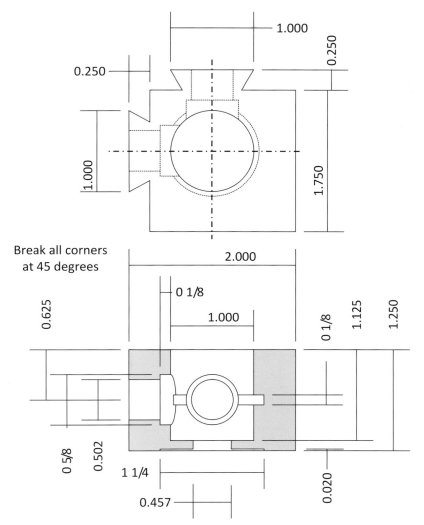

1.000

0.250

0.250

1.000

1.750

Break all corners
at 45 degrees

2.000

0 1/8

1.000

0.625

0 1/8

1.125

1.250

0 5/8

0.502

1 1/4

0.457

0.020

Fig. 20.5 Toolpost block.

To finish the dovetails, use a dovetail cutter, set to cut 0.250in down from the top surface, on one side of a step, cutting in stages until the small 'witness' between the top face and the sloping face just disappears. Zero your DRO or handwheel index at this point. Now move to the opposite face of the step and machine the other dovetail the same way, noting the final distance reading. When you machine the second pair of dovetails, start the same way, zeroing the index or DRO when you finish the first dovetail. When you machine the opposite face of this second dovetail, make sure your finishing cut is at the distance you recorded at the end of the first one. The two pairs of dovetails will now match perfectly, ensuring all your toolholders can be fitted to either of them.

As an additional touch, mill a small pocket on the underside of the block for the catch on the top slide, allowing the block to be reset correctly more rapidly. To add this, scribe a line exactly on the mid-point of the underside of the block. Now mount the block in the mill, inclined backwards at 10 degrees from the vertical. Line up a 1/2in (or similar) end mill with the scribed line and mill the pocket about 1/8in deep. The pocket should be behind the 'front' dovetail and be milled with the side dovetail facing down. If you are unsure, copy the pockets on the original four-way toolpost. Although I have done this on my toolpost, I am not sure how useful it is – the setting is not particularly repeatable, and unlike the four-way toolpost, you should rarely need to turn the quick-release version.

them by boring. The same recessing tool you used to cut the internal groove can also be used to cut the counterbores. The depth of the counterbores is 1/8in at the shallowest point. This is to allow the rim on the pistons to move fully out of the way when fitting the cam.

You now need to cut the two dovetails. (The method described here relies on the block being a reasonably accurate 2in across.) Mount the block upright in a milling vice, and zero a decent-sized slot drill to the top and side of the block. Cutting a full 0.250in depth, take successive cuts towards the centre until you have removed a step 0.500in long. Repeat from the other side of the block. You should be left with a central raised area an inch wide with the piston bore at its centre. Do the same to the face with the other piston bore and check both raised steps are the same width.

Tidy up the block with a fine file to remove any burrs and 'break the arris' on machined edges. You can draw file and then use emery paper to make the unmachined faces look a bit brighter. It is up to you if you want to paint any surfaces of the block.

The Toolpost Clamp Pillar

This item takes the full force of the toolpost clamping nut and transfers it to the bottom of the block. If possible, it should be made of a medium-carbon steel such as EN24T, rather than plain mild steel.

The pillar is drilled clearance for the toolpost stud and is a light push fit in the hole left at the bottom of the recess in the block. Make sure the step on the end is slightly shorter than the thickness of the base of the block, so that when the toolpost clamp is tightened up it grips the block firmly in place. The length of the middle section of the pillar is a critical dimension – it should be the same as the depth of the central recess in the block. Note that this toolpost

uses the existing lathe toolpost clamp (although you could make your own copy if you wish).

It is important that the clamp-screw handle points in the right direction when it is tightened up, so start by making the pillar slightly tall. If the handle points the wrong way, you can face off the end a little (0.3mm off the end will allow the clamp handle to turn about 90 degrees further).

Cam

The cam is a deceptively simple item. There is little benefit in making it from medium-carbon steel, as the wearing surfaces are large and the impact of minor wear on the action of the cam will be negligible. The reference point for all dimensions is its top. The thickness of the upper part should be about 0.002in greater than the depth of the recess in the top. This ensures that when the pillar is locked in place by the toolpost clamp the cam will turn freely without perceptible shake.

The various operations should be done in the following order:

1. In the three-jaw chuck, face off one end of the bar and drill and bore it 0.760in or greater – a very loose fit on the pillar.
2. Turn the outer diameter of the top part and counterbore the recess 0.377in deep.
3. Reverse the bar and trim it to length – 1.490in.
4. Turn the bottom section to a close push fit in the toolpost body, perhaps 0.998in, and so that the length leaves the top part exactly 0.375in thick (in other words, 2 thou less than the depth of the recess).
5. Move the bar to the four-jaw chuck, offset it by 3/64in (this is not a critical dimension), and turn the offset section, noting that its top and bottom edges should match with the piston holes in the block.
6. Next, return the bar to the three-jaw chuck in order to turn the 10-degree taper on the top section (Fig. 20.8).

Do not make the tapped hole for the handle until later, otherwise, like me, you will end up with an extra hole.

Pistons

The pistons can be turned from 5/8in or 15mm brass rod if 9/16in is not available. Turn the main part of the pistons to a sliding fit in the block, making sure there is a good smooth finish. The flanges take no load; they only need to be big enough to stop the pistons falling right through the holes.

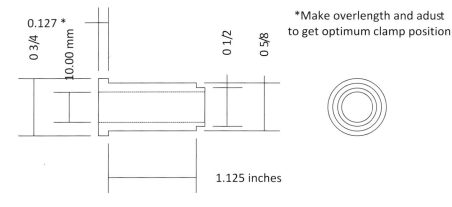

0.127 *

O 3/4

10.00 mm

O 1/2

O 5/8

*Make overlength and adust to get optimum clamp position

1.125 inches

Medium carbon steel

Fig. 20.6 Toolpost clamp pillar.

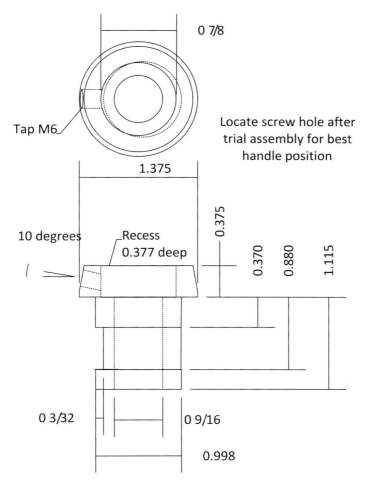

Tap M6

0 7/8

Locate screw hole after trial assembly for best handle position

1.375

10 degrees

Recess 0.377 deep

0.375

0.370

0.880

1.115

0 3/32

0 9/16

0.998

Fig. 20.7 Cam.

Fig. 20.10 Automatic piston retraction.

be fairly accurate with this hole. A good centre pop will ensure the drill does not skid. Clamp the vice to the drill table and, once the hole is started, check the drill is not bending off course. Don't forget to clear the flutes regularly, as small drills clog rapidly. You now need to file flats on the piston flanges, so they don't foul the outer diameter of the cam. These can be done by eye, but you need to ensure that they are parallel with the cross-drilled hole.

Try the pistons in their bores with the cam in place and make sure they are both free-moving and don't interfere with each other, aside from the desired in–out action. The pistons should be just below the surface of the dovetail when fully retracted.

Before final assembly, you should do a trial fitting of a toolholder on each dovetail in turn, to identify the best position for the handle.

(At this point, it is a good idea to make the first toolholder – *see* below – at least as far as cutting the dovetail on the back of a toolholder blank. The rest of the process of making the toolpost will be described first, however.)

Choose the position of the handles so that they do not get in the way of fitting the toolholders when the relevant piston is retracted, and about a 90-degree turn locks the toolholder

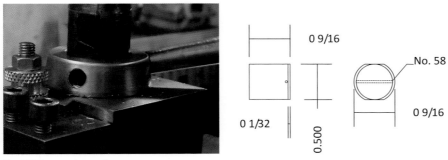

Fig. 20.8 Taper on top section.

0 9/16

0 1/32

0.500

No. 58

0 9/16

Fig. 20.9 Pistons.

So that you can have sprung pistons that withdraw automatically (Fig. 20.10), you need to drill No. 58 (this size is a loose fit for a 1mm diameter spring wire) right through. It is not essential to achieve perfection, but you do need to

without the handle getting in the way of lathe operation. Mark the best position for the hole. Do not assume the best position without this test. If you do, you will end up with an extra hole, as I did. If this happens, you can blank the hole with a short stub of M6 thread, which will virtually disappear if carefully filed and smoothed with emery paper.

The hole for the handle (Fig. 20.11) should ideally be drilled with the cam in an angle vice at 10 degrees. If you do not have one, raising one end of a normal drill vice on a block will suffice – if due care is taken. Check the handle will be in the right place by fitting an ordinary M6 bolt and doing a dry run before fitting the springs.

On final assembly, fit the 1¼in lengths of spring wire in the cross-holes. Make sure everything is clean and free of swarf, then lightly grease the pistons and fit them into their bores, aligning the springs with the groove in the block. Molybdenum-loaded or Teflon grease works well for parts that do not continually rotate. If you grease the cam and line up the bevels on its base with the pistons, it should be possible to insert the cam with a light click as it goes home. It will now be very difficult to take

Fig. 20.11 The handle directions are important.

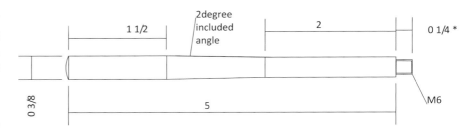

Fig. 20.12 Handle.

*adjust length of thread to avoid interference with toolpost clamp nut

the cam back out and disassemble the toolpost. If you need to do this, drill and tap the pistons, say 8BA, for screws that can be used to pull them forwards so the cam can retract. The flanges will stop the pistons rotating and allow this to be done. Alternatively, if you are super-cautious you can make these holes prior to parting the pistons off.

Handle

The toolpost is now complete, aside from the handle. This is simply made from a length of 3/8in bar, with an M6 thread on one end. The tapered section should be turned at about 1-degree (2-degree included) angle to match the standard mini-lathe handles. The handle can now be screwed into place, the pillar popped into the centre hole of the cam, and the toolpost fitted to the lathe.

TOOLHOLDERS

The basic toolholder is a 2in length of 1in square mild steel. Each one needs a dovetail cutting in the back. Make a template from aluminium sheet that is a good fit on the toolpost block dovetails, and mark it with two lines 50mm (2in) apart. This will allow a quick check

that the dovetails are properly placed and proportioned (Fig. 20.14).

If you have a decent length of steel available, it is a good idea to 'mass produce' a quantity of blanks – you should get nine from about 20in of bar – carrying out one operation at a time. Start off by milling a 1/4in deep slot in the middle of each blank 0.710in wide (Fig. 20.15). This is the width across the narrow base of the dovetails. Use a good-sized slot drill (I used 11.5mm) to ensure you get a consistent width. Aim to get the slot in the centre of the blank – a rule dimension or measuring from the template is fine. The depth should be at least 0.250in and reasonably accurate. Too deep means the lever needs to turn too far, which may be inconvenient; keep the same depth for each toolholder, so the handle always locks in the same place. Similarly, the width should be as accurate as possible in order to facilitate accurately cutting the dovetail itself.

The procedure is as follows:

1. Cut one side of the slot in two passes to exactly 0.250in deep.
2. Take further cuts at the full 1/4in depth until the slot is full width. A

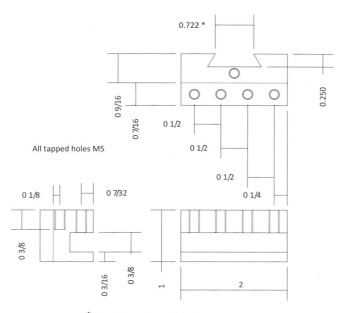

All tapped holes M5

*Adjust dovetail to match toolpost
dovetail detail same for all styles of holder

Fig. 20.13 Standard toolholder.

Fig. 20.16 Blank with full dovetail.

Fig. 20.14 Dovetail template.

Fig. 20.15 Blank with rectangular groove.

digital readout will facilitate this; simply subtract the cutter width from the slot width, and zero the readout when cutting the first side.

3. Once you have one blank slotted, ditto repeato (to coin a phrase) until you have a pile of prepared blanks.

4. Now exchange your slot drill for a dovetail cutter. For each blank, start by adjusting the cutter to 0.250in depth. Take suitable passes along one side of the slot until the

witness of the vertical edge of the slot disappears.

5. Zero the digital readout (or your handwheel index).

6. Next, repeat on the opposite edge of the slot – you should find that just removing the witness on the second edge creates a dovetailed slot that is just a bit too tight. Gradually widen the slot until the dovetail is an easy fit on the toolpost block, ensuring that the dovetails are clear of swarf

before testing the fit. Make this test without removing the toolholder blank from the mill so you do not lose the index setting. Once you have a suitable fit, note the DRO or index reading.

7. You can now machine the dovetails on the remaining blanks (Fig. 20.16), working to the same reading, in confidence that they should all be a good fit on the toolholder. (In practice, I tested the fit 10 thou before reaching the 'ideal width', to be on the safe side.)

8. With a suitable fine file, tidy up the edges of all the blanks and make sure they are all a good fit on the toolpost.

Height-Setting Studs and Buttons

These studs and their nuts or buttons are what give the quick-release toolpost its repeatability. Drill and tap the blind M5 holes for the studs in the blanks. Note how close they are to the dovetails. This ensures a good overlap of the buttons and the toolpost and minimizes interference between the buttons and tool fixing screws. The

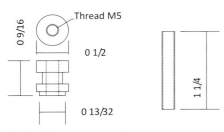

Fig. 20.17 Height-setting studs and buttons.

Fig. 20.19 Parting off buttons.

Fig. 20.20 HSS tool in standard toolholder.

studs themselves are just lengths of M5 studding.

The height-setting buttons can be finished to suit your own preferences. All that matters is that they have a suitable flat base that will overlap the toolpost (Fig. 20.18). To make the height-setting nuts, I made a diamond knurl along just over half of a length of 1/2in brass bar. I then reversed it and held it in the chuck, using some aluminium shim to prevent damage to the first section, and then knurled the remainder of the bar. I used a 1/8in parting tool to create the decorative effects, repeated along the bar, and then a 1/16in parting tool to cut off the individual nuts (Fig. 20.19).

Again, using the aluminium shim for protection, I faced off the pip left by parting and then drilled and tapped

Fig. 20.18 Stud and button in place.

each nut M5. I used ordinary M5 nuts to lock these height-setting nuts; you may wish to make a batch of smaller brass lock nuts.

CUSTOMIZING TOOLHOLDERS

It is now up to you to decide what different types of toolholder you need and how many of each type. Take your time and enjoy the variety of tasks after all that repetition work!

Standard Toolholders

A standard-size high-speed steel (HSS) tool for smaller lathes is typically 8mm or 5/16in HSS. The standard toolholder for these bits requires a suitable slot to be milled along the length of the holder (as in the drawings) and threaded holes to be made for four holding screws. If you can get M5 grub screws, these are a neat solution, but if not Screwfix and other suppliers offer 16mm stainless steel cap screws at a very reasonable price for 50. Some have the advantage of ends that are turned neatly and in such a way that they will not burr over.

These holders can also be used for 1/4in HSS, but if you prefer this size you may find you can economize and make

toolholders from 3/4in bar instead. If so, you will need to use grub screws to make sure they do not foul the height-setting nuts.

The ends of the toolholders can be treated in whatever way you wish. I plan to bevel mine at about 30 degrees for the sake of appearance only – it will not materially affect their strength.

Boring Bar

For a boring bar holder you need to drill and, ideally, ream a hole that is a good fit for your boring bar along the length of the holder. I prefer two or three screws bearing on the bar to hold it in position. Others may prefer to slot the holder and make a clamp arrangement. Ideally, the hole for the boring bar should be reamed to size. I drilled undersize and used a hand reamer to finish the hole.

Be careful not to overtighten the clamp screws, especially if your boring bar is unhardened steel, as you could damage its surface, making it difficult to remove. If this is a concern, use a clamp arrangement instead, or put brass or copper pads down the screw holes. I have a long inserted cutter bar and several smaller HSS boring bars with 3/8in diameter. The former is hardened

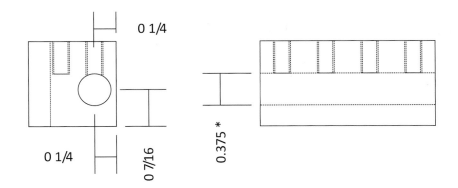

0 1/4

0 1/4

0 7/16

0.375 *

*ream bore as required to suit other tools
Other details as standard toolholder

Fig. 20.21 Boring bar holder.

Fig. 20.22 Boring bar fitted in holder.

steel and the latter are HSS, so I am not worried about this issue.

I made two of these holders – the first has a 0.375in hole that fits both my larger boring bars (Fig. 20.22), and the other is bored 0.25in. For the smaller one, I have also made a sleeve for a very small 3/16in bar.

Parting Tool

There are several different styles of parting blade available, some of which are rectangular in section, while others have some relief on the sides. I made my holder (Fig. 20.23) to suit 1/16in by 5/16in section HSS. This shape allows

the use of a flat-sided holding groove, but you will still need to obtain or make a small dovetail cutter to form the bottom edge of the groove. Such a cutter can be made from 3/8in silver steel by turning a short cone on the end, filing a few teeth and hardening and tempering it. Such a simple tool will be fine for this task, although it will be a bit crude for making proper dovetails.

Non-rectangular blades will require a recess in the block to match their cross section, to ensure they sit vertically in the holder.

The clamping block, shown in Figs 20.24 and 20.25, only extends for half the length of the groove for two reasons: it concentrates the clamping force where it is needed, and it gives it a more positive location. I suggest fitting the block in place before drilling tapping size for the clamp screw right through both components.

DTI Holder

As a dial test indicator (DTI) is not under heavy loads (hopefully!), it can be made from steel or a length of 1/2in thick aluminium alloy. The drawing (Fig. 20.26) is almost redundant – this is simply an over-length toolholder with a 3/8in hole (or whatever suits your DTI) and a clamp screw. This is another opportunity to have a bit of fun, and see if you can make a nice shape, rather than just producing something blandly functional. Do think about making the clamp screw from brass with a knurled knob, to make sure it does not get overtightened. The right-handed version in the drawing is suitable for holding a DTI to check the concentricity of work

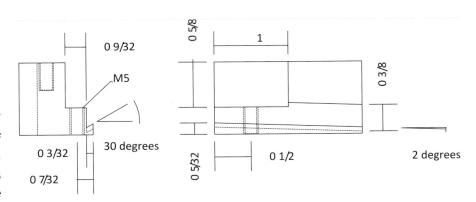

0 9/32

M5

0 3/32

0 7/32

30 degrees

0 5/8

1

0 3/8

0 5/32

0 1/2

2 degrees

Other details as standard toolholder

Fig. 20.23 Parting toolholder.

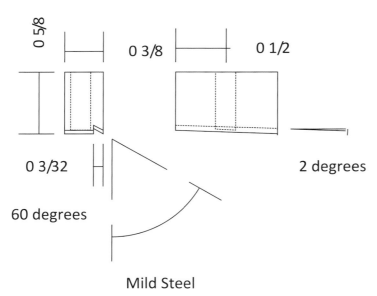

0 5/8

0 3/8

0 1/2

0 3/32

2 degrees

60 degrees

Mild Steel

Fig. 20.24 Clamping block.

Fig. 20.25 Parting tool fitted to holder.

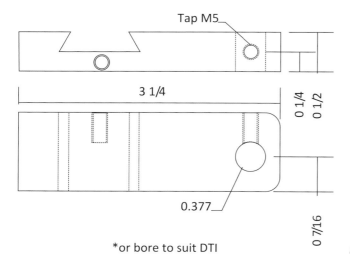

Tap M5

3 1/4

0 1/4

0 1/2

0.377

*or bore to suit DTI

0 7/16

Fig. 20.26 DTI holder.

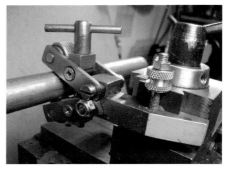

Fig. 20.27 Knurling tool adapted to fit quick-change toolpost.

held in a four-jaw chuck. A left-handed version could be modified to support a 'Verdict'-type lever indicator for checking the truth inside a bore.

Knurling Tool

One of the very first workshop tools I made was an unsophisticated, but effective, clamp knurling tool. This was made with a bar that clamped in the four-way toolpost. I simply cut a dovetail in the bar and added a height-setting stud. Although this has not made the tool any prettier, it works well (I knurled a long length of 1/2in diameter brass with this device to make the height-setting buttons) and demonstrates the principle that any tool that can be fixed in the toolpost can potentially be modified in this way.

A Few More Thoughts...

There are more ideas to consider: a holder reamed for an MT1 taper, my keyway slotting tool to fit the toolpost, or a modified version of my toolpost drill. Equally, a toolpost grinder, multitool or any other device could be modified to fit on the dovetails. There are probably many other possibilities too.

Many readers will know how popular tangential toolholders are. I have made one of these for 1/4in HSS and it has proved very useful. A quick-release toolholder version is shown in the concept drawing here – no dimensions are given, as the design would almost certainly benefit from some adjustments. This is something I have not seen elsewhere,

Fig. 20.28 Tangential toolholder.

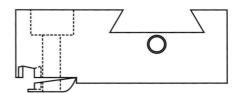

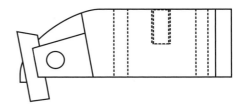

Fig. 20.29 Concept for tangential toolholder.

but it seems blindingly obvious. Note that, as the toolbit can be adjusted separately for height, the holder needs to be set as low as possible and maximize the amount of metal supporting the toolbit. This is why the lower part of the dovetailed body will need to be cut away.

As it is now so easy to adjust the height of your tools to be spot on with this new toolpost, the value of a height-setting gauge becomes apparent! It also means there is no excuse not to swap tools as often as you need, rather than making do with the same tool for turning and facing, or accepting the finish from a roughing tool.

21 Looking After Your Tools

It is good practice always to keep your tools clean and well organized. While a hammer obviously requires less protection than a delicate micrometer, all tools benefit from protection from knocks and rattling around with other tools. At the same time, organized storage makes things easier and quicker to access.

Your most frequently used tools might live on the workbench itself, lie on handily placed shelves or be hung on the walls. Tools and tooling that are used less often will tend to live in drawers, while the most delicate or expensive will be kept inside storage boxes. My personal preference is for all my most frequently used items to be kept on shelves, so that they are easily to hand. Despite a superficially chaotic arrangement, almost everything will have a specific place.

STORAGE UNITS, BOXES AND DRAWERS

A good friend of mine used to have a small but unusual collection of the toolboxes of departed engineers he had known. They varied in the detail, but each was a wooden cabinet with a number of small drawers. Many of them still contained part of their previous owner's collection of precious tools. It is still possible to buy expertly made engineers' toolboxes of this kind, or to make your own. Some are relatively small and portable; others are tall units designed to be kept in place on a workbench. Many are very workmanlike in design, while others display expert crafting in their construction, with hidden dovetails and polished walnut. The drawers are often quite shallow, and should be made of a type of wood that will not corrode steel – oak, for example, is notorious for encouraging rust. Sometimes, the drawers are lined with felt. Portable versions have a large compartment on top with a hinged lid, typically with a mirror on its underside – opinions vary on the use for this! They are ideal for storing relatively delicate tools, especially gauges and measuring tools.

The more modern equivalents are steel cabinets with ball-bearing drawers, from modest toolboxes to units on castors the size of a fridge. These tend to be rather larger and are a ubiquitous choice for garages and similar establishments.

For those working on a budget there are several options for drawers. It is always worth looking around second-hand furniture shops and boot sales. Modern units tend to have very large, relatively flimsy drawers and are best suited to bulky, but not weighty, items. Older dressing tables and tallboys often have shallower drawers made of solid timber and plywood and can be ideal for storing heavier tools. Do be selective, though – better-quality units in solid oak are not a good choice!

Fig. 21.1 Lathe tooling supplied in wooden box.

Fig. 21.2 A set of repurposed drawers.

Fig. 21.3 Plastic storage drawers.

While some tools can survive a degree of rough and tumble, anything with a precision function, delicate mechanism or cutting edge, for example, needs to be treated with more respect than simply being thrown in a drawer. Reamers and files are a good example of tools that will last much longer if they are kept from knocking into each other. Tools can be protected with a length of pipe around a vulnerable part, or by being slipped into a plastic wallet. For the most delicate tools, the best solution is a dedicated box, but for many tools dividing drawers to stop things rolling around will be sufficient. Dividers are easily made from pieces of timber or fibreboard, or you can arrange card or plastic boxes within a drawer. Cutlery drawer inserts are an easy way to store small tools in a large drawer.

Soft drawer liners can also help reduce movement and protect tools. Also it is the traditional choice, felt tends to pick

If you want to make your own storage – say, several shallow, strong drawers – wheeled and ball-bearing drawer runners are quite inexpensive. You can use them to construct units from plywood-bottomed boxes running within a simple carcass of sheet material or even an open wooden frame.

For under-bench and benchtop storage of small tools, plastic drawers meant for home office use can be convenient. Make sure that they open easily and are strong enough for the job – some are not very robust.

If you want small wooden storage units, look in local craft shops for unpainted drawer sets and trinket boxes. These are often in softer timbers but, if treated with care, they can be long-lived. You can also repurpose boxes intended for selections of artists' paints and brushes for storing thin, flat measuring tools.

These days, many tools come supplied in good-quality boxes, however the 'blown plastic' types are often too bulky for convenience. Often a better box can be found in the endless supply of cardboard boxes that enclose everything we buy, and the plastic one can go for recycling.

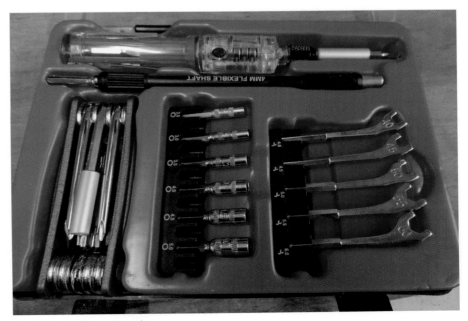

Fig. 21.4 Space-efficient storage box.

Fig. 21.6 Workshop thermometer and hygrometer.

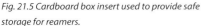

Fig. 21.5 Cardboard box insert used to provide safe storage for reamers.

up swarf and oil. A better alternative is an open mesh of synthetic rubber, which can stop tools skidding around, even inside metal drawers, and is easy to clean. It also makes metal drawers rather quieter in use.

DAMP

Dampness is the biggest single destroyer of tools and materials, with rust being the chief issue. If your workshop is truly dry, you may never experience rust as a problem, but others may find themselves continually battling against it. There are also workshops where damp causes issues only on rare occasions. A maximum/minimum humidity and temperature meter is an inexpensive way to monitor what is happening in the atmosphere and to get an understanding of the problem. Such a meter also provides reassurance that your precautions against damp are being effective.

Leaks are obviously to be avoided, but the real problem is condensation, which is primarily caused by rapid changes in temperature. The obvious solution is to insulate the workshop and prevent draughts as much as you can, just as you would in your home. Consider some form of low-level heating to avoid frost. My own workshop has a metal roof and without insulation it would drip pints of water from a single frosty evening.

Another tip is to avoid storing things in ways that allow damp to build up against walls or in corners. Try to allow a free flow of air wherever you can.

If occasional rusting of tools left on a bench or shelf is an issue, it may be sufficient to keep them inside dry drawers. Sometimes, it is possible to convert shelves into cupboards, just by adding doors. For long-term storage, use sealed plastic boxes; it can be convenient to buy large boxes and place several smaller ones inside.

If, despite these measures, you still have problems with rust and damp, there are a number of steps you can take:

◆ Dehumidifiers have become affordable both to buy and to run. One

that has a good strong fan will help circulate air around the workshop as well as keeping the humidity down. A small dehumidifier will use only a small amount of energy and provide a source of background heat to reduce the risk of frost. Choose one with a humistat, so that it does not waste money running when the air is dry. For woodworkers, a humidity of around 45 per cent is often advised, but for metalworking workshops it can be much higher than this, while still protecting tools and stored materials effectively. It is also a good idea to choose a dehumidifier that has an external drain, so you do not have to keep emptying it.

◆ Dehumidifiers that use hygroscopic salts can be used to keep drawers and boxes dry. 3D printer filament stored in a large, sealed storage box alongside such a dehumidifier should remain flexible and reliable for a year or more.

◆ Rust inhibitor paper is a surprisingly inexpensive but effective material that gives off chemicals that form a thin protective coating on bare metal. For long-term storage, placing slips of rust inhibitor paper alongside the tools is remarkably effective.

- ◆ Rust-proof films offer a more direct approach that is particularly useful for items such as the beds of machine tools, but can also be used for small tools. Just wipe on a very thin coating – the film will be so thin that it will not affect the accuracy of gauges and measuring tools.

- ◆ For larger tools such as chisels, water-dispersing oils and protective waxes can be applied to make a more robust and long-lived film.

Fig. 21.8 A permanent drain removes the need to empty the dehumidifier.

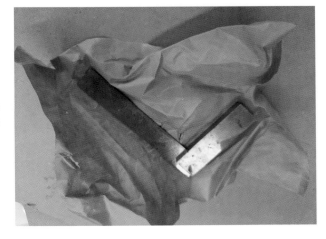

Fig. 21.9 Rust inhibitor paper is very effective.

Fig. 21.7 A small dehumidifier consumes very little energy.

My aim in writing this book has been to give you useful advice on the care and use of smaller tools, and inspiration for making your own tooling, from specialist cutters to some fairly complex devices. Above all, I hope it will add to the satisfaction and enjoyment you get from a well-equipped workshop.

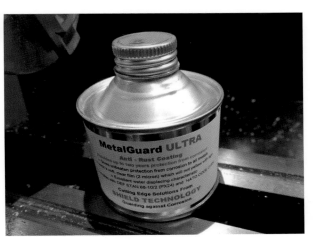

Fig. 21.10 Rust inhibitor films can be used on hand tools and machine tools.

Index

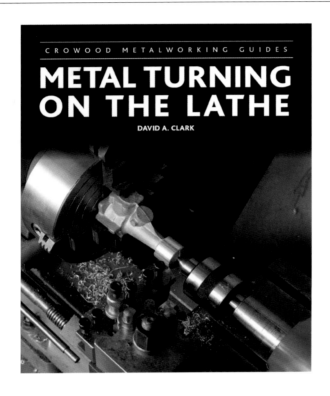

CROWOOD METALWORKING GUIDES

METAL TURNING ON THE LATHE

DAVID A. CLARK

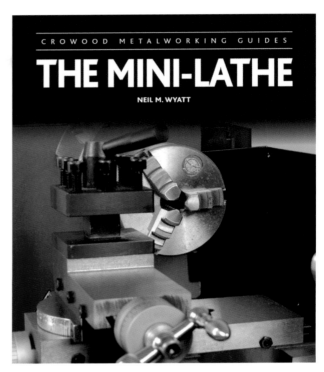

CROWOOD METALWORKING GUIDES

THE MINI-LATHE

NEIL M. WYATT

CROWOOD METALWORKING GUIDES

SHARPENING
COMMON WORKSHOP TOOLS

DR MARCUS BOWMAN

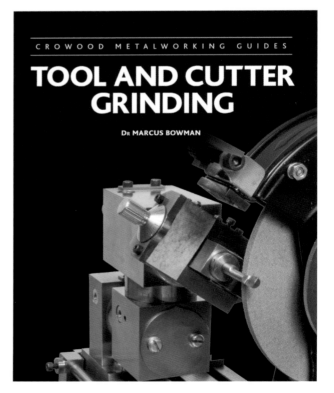

CROWOOD METALWORKING GUIDES

TOOL AND CUTTER GRINDING

DR MARCUS BOWMAN